Coexistence and Coevolution of Humans and Intelligent, Adaptive Environments

Humankind is now in an age wherein generations of humans have never lived without ubiquitous computing and the internet. This AI/IoT generation is creative, innovative, and adaptable; however, the relentlessness, complexity, and volume of technological, biotechnological, healthcare, social, cultural, and emotional changes and stressors cause humans many personal, social, and environmental traumas. This book presents a vision for designing and managing the complexity and multidimensionality of the world's emerging technological and AI/robotics-enabled environments that are grounded in current methods for designing human-centered, complex, interactive systems.

This book discusses the coexistence and coevolution of humans and intelligent, adaptive environments during a transformational age during which society is confronting social, environmental, and health crises. Defining the creative strategies needed as "heroes" of the near future, this book investigates the coexistence/coevolution concept that entails how humans should coevolve with their "smart artifact ecologies" by designing dynamic environments that ameliorate technology-induced trauma. In understanding the emerging AI-enabled, Non-Humanoid Social Robotics (NH-SR) and IoT-enabled environments, the reader will develop an understanding of how to design environments for future generations.

Coexistence and Coevolution of Humans and Intelligent, Adaptive Environments is a thoughtful read for graduate students, researchers, and academics interested in Architecture, Systems Engineering, Computational/Robotic/Intelligent Architecture, Human Factor Engineering, Robotic Engineering, Mechatronics Engineering, Embodied Cognition, Cognitive Psychology, Human-Systems Integration, and Human-AI-Robot Teaming. It equally speaks to social scientists, humanists, futurists, and visionary architects whose work navigates the evolving entanglements of technology, humanity, and design.

Coexistence and Coevolution of Humans and Intelligent, Adaptive Environments

Tarek H. Mokhtar
Joseph Manganelli

CRC Press
Taylor & Francis Group
Boca Raton London New York

CRC Press is an imprint of the
Taylor & Francis Group, an **informa** business

Designed cover image: Tarek H. Mokhtar

First edition published 2026
by CRC Press
2385 NW Executive Center Drive, Suite 320, Boca Raton FL 33431

and by CRC Press
4 Park Square, Milton Park, Abingdon, Oxon, OX14 4RN

CRC Press is an imprint of Taylor & Francis Group, LLC

© 2026 Tarek H. Mokhtar and Joseph Manganelli

First edition published by CRC Press 2026

ISBN: 978-1-032-57990-0 (hbk)
ISBN: 978-1-032-56882-9 (pbk)
ISBN: 978-1-003-44195-3 (ebk)

DOI: 10.1201/9781003441953

Typeset in Times
by SPi Technologies India Pvt Ltd (Straive)

Dedication

From Tarek Mokhtar

To my wife Sara Sahab, whose unwavering patience sustained me during research and writing. This book carries her quiet strength on every page. To my sons, Youssef Mokhtar, Amir Mokhtar, and Adam Mokhtar, whose laughter and curiosity remind me why futures worth designing matter. To my parents, Enas Eldessouki and Hassan Mokhtar, and my sister, Randa Mokhtar, whose unwavering faith and gentle guidance remain my foundation. I extend special thanks to Saudi Arabia's Research Development and Innovation Authority for their generous support (Grant #: 12947-ALFAISAL-2023-FU-R-3-1-EF).

From Joseph Manganelli

To my wife, son, parents, and sisters, thank you for your support during these last few years.

Contents

About the Authors xi
Preface xii
Acknowledgments xiv

Introduction 1
    The Fundamental Question 1
    Developing the Answer to the Fundamental Question 1
    The Time Period 1
    The Set of Places 2
    The Set of Users 3
    Factors Influencing the Answer to the Fundamental Question 3
    Rays of Hope 5
    The Answer to the Fundamental Question 5
    "A-Day-in-the-Life-of Humans"...2050 Scenarios with Paulette
    Romilly 7
    Day in Life Scenario in 2050 7
    From Trauma, Sufferings, and Colonizers to Rays of Hope and the
    All-in-One Environment 9
        Part I: Colonizations at the Age of Transformations 10
        Part II: Heroes of the Past and the Present 10
        Part III: Sufferings of Humans in the 21st Century 11
        Part IV: Rays of Hope/Heroes of the Near Future 11
        Part V: Designing Coexistence/Coevolution 11
        Part VI: Epilogue: All-In-One (AIO) Environments 11

Part I   Colonizations at the Age of Transformations 12
    Colonization... 12
    In Transformation 15
        Poly-Colonization: An Expanded Definition 15
        Philosophical Perspectives on Poly-Colonization 16
        ...Parasuraman: Neuroergonomics 17
        ...On Trust and Distrust 18
        ...On Language 18
        ...Decentralized Nature of Power 19
        ...On Identity and Agency 19
        ...In Art 20
        ...On Trauma and the Self 20
        ...On Architecture and the Built Environment 21
    At the Age of Transformation 21
        Colonizer I: Pandemics and Viruses 22

Colonizer II: Genetic Developments 23
Colonizer III: Digital and Cyber Colonization 24
Colonizer IV: Health Colonization 25
Colonizer V: Environmental and Ecological Impact 27
Stressors and Humans-Technology-Environment.....................29
Traumas Caused by Poly-Colonization (for Technologists,
Architects, and Every Human on Earth—Tables Awaiting
Your Response)..30
The Threshold of Tomorrow 30

Part II   Heroes of the Past and the Present..38

Heroes in Place 38
Heroes...Defeating Mechanisms For......................................40
...Pandemics and Viruses...41
...Genetic Developments...42
...Digital and Cyber Colonization 42
...Health Colonization 44
...Environmental and Ecological Impact................................45
Criticism of Our Defeating Mechanisms 46
...On Design and Technology..47
Evolving Classical Programs of Architectural Systems 49
Limits of Adapting to New Tools and Systems 49
Limits of Predictive Analytics for Human Performance and
Well-Being...50
The Unpredictability of Social Trends....................................50
Ecological Niche Construction 51
The Limits of Our Heroes… 52

Part III Sufferings of Humans in the 21st Century............................53

Contemporary Sufferings...53
Humans and the Digital Dilemma...55
Humans and Environmental Crises...Physical and Mental Health........59
The Health Paradox...64
The Economy of Attention..68
Technological Unemployment and the Skills Gap...................73
Social Fragmentation and Polarization...................................78
The Future of Work...81
Humans, Architecture, and the Digital Dilemma....................85

Part IV Rays of Hope/Heroes of the Near Future...............................87

Portal to the Future...87
Building Blocks… 90
Ray I: Automated Systems..90
Ray II: Human Systems (Science of Well-Being) 97

Ray III: Systems Validation 99
Ray IV: Embodied Action and Cognition 106
…Human-in-the-Loop 110
Ray V: Human-in-the-Loop Automated Systems 110
…Systems 113
Ray VI: Human-Built Environmental Systems....................... 113
Ray VII: Non-Humanoid Social Robotics (NH-SR):
Redefining Interaction 117
Looking Ahead…Heroes of the Near Future 122

Part V   Designing Coexistence/Coevolution 123

On Symbiosis… 123
Symbiosis from Biology to Building 125
Levels of Symbiosis in Architecture 125
Individual Autonomy and Co-evolution in Buildings.............. 126
Community-Scale Symbiosis: ULSS and Ecological
Integration 126
Societal Scale, Macro Perspectives, and Ethical
Implications 127
Toward Participatory and Adaptive Futures............................ 129
Redefining Interaction | Transcending Boundaries… 130
Symbiotic Futures… 133
Symbiotic Future I: "Social Dialogues"…A Balance
between Social Needs and Technical Capabilities.................. 133
Symbiotic Future II: "Healing Environments"…Healing
Through Innovation 139
Symbiotic Future III: "Evolving Complexity"…
Complexity of Needs 145
Artificial Symbiosis …Evolution and Adaptation… 148
Ecological Niche Construction in Architectural Design.......... 149
Ecological Niche Construction in Modern Cognition and
Technology... 149
Ecological Niche Construction in the Age of Cognitive
Architecture and Technological Integration..................... 150

Part VI   Epilogue: All-in-One (AIO) Environments........................... 151

For Paulette… 151
AIO Environments 153
Defining the AIO Paradigm 153
Principles, Features, and Functions 154
Communication and Multimodal Feedback........................... 160
Dynamic Environmental States.................................... 161
Non-Humanoid Social Robotics 161
Ethics of Trust and Agency in Intelligent Spaces 162

Artificial Symbiosis in AIO Environments 164
   Paulette in AIOs for Engagement 165
   Paulette in AIOs for Learning.. 167
   Paulette in AIOs for Health.. 168
   Paulette in AIOs for Work.. 169
   AIO as a Manifestation of Symbiosis.................................... 170
Paulette + AIOs…Coexistence and Coevolution… 171

Bibliography 175
Index.. 196

About the Authors

Tarek H. Mokhtar, Assoc. AIA, RA-EG, PhD, MSc, is a prominent scholar, architect, and Assistant Professor of Architectural Engineering who moves at the interstice of art, philosophy, robotics and machine intelligence, crafting works that merge human needs with philosophical, social, and artistic provocation. Directing the Dimensions Startup Studios and the DaVinciAT Robotics Research Center at Alfaisal University, he leads research in cyber-physical systems, human–robot interaction, and intelligent responsive environments—crafting a future where the spatial, the social, and the computational converge into a living architecture. For over two decades, his teaching, research, and practice have pursued a singular premise: that the built environment must not only shelter, but comprehend the shifting emotional and social ecologies of its inhabitants. From his Master's thesis on the New Scenario of Aesthetics: Philosophy and Meaning of Beauty in Millennium Architecture to his doctoral dissertation, of Clemson University, USA, "Monumental-IT: A 'Robotic-Wiki' Monument for Embodied Interaction in the Information World," he has advanced Architectural Robotics and coined Non-Humanoid Social Robotics on his well-recognized Springer book chapter, framing space as a social agent capable of trust, empathy, and interaction without anthropomorphic mimicry.

Mokhtar's architecture blends art, technology, and human experience into systems that listen, learn, and heal. His U.S. Patent for aiding ARDS patients embodies his conviction that design, when fused with empathy and intelligence, can restore dignity and agency. As an artist, he creates artworks where the tactile meets the algorithmic, inviting reflection on the nature of human–machine coexistence.

Joseph Manganelli, AIA, CHFP, LEED AP BD+C, PhD, is a practicing registered architect, board-certified human factors professional, and a sustainable design specialist. His goal is improving the capacity of the built environment to enhance human cognitive and physical health, well-being, and performance. Dr. Manganelli's research and practice integrates constructs, measures, and methods from architecture, human factors, systems engineering, cognitive science, evolutionary biology, and sustainability. His architectural practice focuses on industrial facilities design and his human factors practice focuses on industrial, healthcare, automotive, and AI concerns. This background was developed to better design for humans in task-oriented, mission-critical environments. It's also led to research creating constructs, measures, and methods for developing human-centered cyber-physical systems and human-AI-robot teams. Dr. Manganelli is Principal Architect and Human Factors Consultant at Fluor, Inc., Part-Time Faculty in Kent State University's School of Information, and owner of xplr design, llc.

Preface

The rate of change, scope of change, and complexity of change of information, tools, and environments affect people's sensing, perception, cognition, and well-being. This is and always has been true, whether the information is about a shift in wind direction or an update to a computer GUI, whether the tool is a rock or an assistive AI agent, and whether the environment is a cave or the audio-visual experience of a rock concert. If the rate of change is too infrequent, and/or the scope of change is too narrow, and/or the complexity of change is too low, then people struggle to maintain vigilance in attending to their situations and lose the capacity to respond effectively when there is substantial change. That is, their performance, adaptability, and resilience gradually degrade, or worse, atrophy, due to lack of use. Conversely, if the rate of change is too rapid, and/or the scope of change is too great, and/or the complexity of change is too high, then people's sensing, perceptual, and cognitive systems are overwhelmed and performance, adaptability, and resilience degrade, even to the point of failure. That is, these systems eventually get damaged or fatigued, cannot function in a healthy way, and damage can compound and become irreversible if the systems are not afforded needed rest time to recover.

In between these two extremes, there is an optimized *range* of rate of change for people's use of information, tools, and environments. *From this perspective, the design and operations goal must be to tune the composition and use of information systems, tools, and environments to achieve a salubrious (not too little, not too much) rate of change, scope of change, and complexity of change, such that they invigorate human sensing, perception, cognition, and well-being—and foster optimization of performance, adaptability, and resilience.* The desired outcome is people who feel comfortable and safe in their information systems, with their tools, in their environments, and who have the agency and capacity to act to maintain this empowered state, individually and collectively.

Achieving this goal is further complicated by many factors. This book addresses three specific sets of complicating factors:

1. Trauma suffered by people in response to socioeconomic, healthcare, and technology stressors;
2. Real-time interactivity with intelligent, adaptive systems;
3. The ethics of designing and operating intelligent, adaptive systems that have the capacity for real-time interactivity.

Regarding the first set of complicating factors, this book discusses stress and trauma that people suffer in response to contemporary socioeconomic, healthcare, and technology stressors as a pervasive, technology-centric colonization, referred to as poly-colonization. Poly-colonization describes how technologies that ostensibly exist to improve our socioeconomic conditions, our healthcare, and our task execution also have the unintended (or intended) side effect of constraining our perceptions, decisions, and behaviors, while also distorting our senses of reality, in ways that may be against our own interests. This creates stress because people are aware that they are being controlled and manipulated but feel powerless to break the bonds of poly-colonization.

At the same time, the advent of real-time interactive systems with advanced, multi-layered, integrated control systems (i.e., autonomy) presents qualitatively different design and operations challenges. Whereas previously, system intelligence was embedded in more or less static structures of information, tools, and environments (i.e., the forms could be static because the behaviors they afforded changed very slowly), now system intelligence can react and evolve in real time as task execution dynamics changes more frequently (i.e., the forms must accommodate behaviors that evolve rapidly and continuously over shorter time scales). This necessitates humans and their information systems, tools, and environments to be more adaptive over shorter durations of time. *How should we design and operate such intelligent, adaptive systems?*

Lastly, the ethics of designing intelligent, adaptive systems that don't exacerbate stressors and traumas but rather ameliorate them is addressed. At a basic level, ethics relate to whether use of a particular technology—or how it is implemented—is detrimental or beneficial for users. At a higher level, ethics relate to how systems of coevolving agents and technologies achieve personal benefits through mutually beneficial symbiotic relationships.

This book presents a summary of likely constructs and methods required for designing and operating intelligent, adaptive environments, as well as a vision for how these intelligent, adaptive environments will fit into our lives—at rest, at work, and at play. This vision frames our information systems, our tools, and our environments as extensions of our minds that coevolve with us. This book advocates a holistic approach to design, grounded in an embodied cognition perspective.

Acknowledgments

For more than a decade, this book has been a convergence of questions, dialogues, and convictions shared between *Tarek Mokhtar, Joseph Manganelli*, and *numerous scientists, researchers, artists, colleagues, friends, and family members*. It would not have been possible without the quiet architectures of support and patience that surrounded us during its making.

Together, we acknowledge the countless conversations—spoken and silent—with colleagues, students, and thinkers who challenged and enriched our inquiry into the entanglement of humans and adaptive environments. This book is not an ending but an invitation to continue that dialogue. The cover, *Coexistence and Coevolution*, designed and illustrated by Tarek Mokhtar, is less an art illustration than an opening argument—a silent preface where line, color, and void begin the dialogue that the text continues. In its creation, no division was sought between the intellectual and the visual, between thinking and making; the cover is the book's first sentence, whispered rather than written.

Introduction

Design of the built environment is and has always been an ethical act because *it can enhance or detract from people's physical and cognitive performance, their social interactions, and their well-being.*

However, the means and methods of designing built environments are changing. They have entered a liminal period, during which the familiar logics of industrial modernity are giving way to a new era of pervasive, AI-enabled and robot-enabled, real-time-adaptable environmental systems. During this transitional era, neither nostalgia for inherited environmental design processes, nor piecemeal optimization of new environmental design processes are adequate. Instead, a paradigmatic shift is required for how the built environment is designed (Brynjolfsson & McAfee, 2014; Manganelli, 2013; Mokhtar, 2011; Mokhtar, 2019).

THE FUNDAMENTAL QUESTION

How do we ethically design intelligent, adaptive environments, imbued with Industry 6.0 technology, that actively support human performance and well-being?

DEVELOPING THE ANSWER TO THE FUNDAMENTAL QUESTION

The answer to the fundamental question can only be defined relative to the dynamics *that coexist during a particular time period, in a particular set of places, for a particular set of users.* Thus, to answer this fundamental question, we must define the time and place for which—and intended users for whom—the answer is valid.

THE TIME PERIOD

This book answers the fundamental question during the transition from Industry 3.0 technology to Industry 6.0 technology (Das & Pan, 2022). That is, the time period addressed in this book is the transition from:

- *Industry 3.0*: buildings enhanced by electrification, mechanical ventilation, plumbing, fire suppression, and telecommunications systems with basic mechano-electrical control systems, and overlaid with a patchwork of basic computer-based automation and control systems (i.e., programmable thermostat), to…
- *Industry 4.0*: Industry 3.0 technology *plus* the incorporation of internet-of-things-enabled (IoT) technologies, cloud computing, basic mechatronics and robotics, advanced digital control systems, and advanced analytics, to…

DOI: 10.1201/9781003441953-1

- *Industry 5.0*: Industry 4.0 technology *plus* the incorporation of human-centered design and operations strategies, which recognize that fitting task execution to human operator abilities optimizes the processes overall while providing better experiences for the humans, to…
- *Industry 6.0*: Industry 5.0 technology *plus* the incorporation of comprehensive sensor networks, perceptual systems, cognitive systems, responsive actuation, *and systems integration and management developed to the point of maturation wherein it is possible for the system of systems to self-regulate and self-optimize in real time in response to human behavior and process changes.*

These transitions are occurring in a heavily time-compressed way (just a few decades spanning the transition across multiple technological epochs). The transitions are so time-compressed that they are practically happening at the same time. Most built environments are still making the transition from Industry 3.0 to Industry 4.0 technology, while a sizable minority are transitioning from Industry 4.0 to Industry 5.0 technology, and only a small number are already transitioning from Industry 5.0 to Industry 6.0 technology.

These transitions are also occurring against a backdrop of Globalism, in which complex industrial, social, and economic networks span communities *across the world*. That is, lived experience is increasingly similar, dependent on socioeconomic status and access to technology, regardless of where one lives, due to global socioeconomic stratification. This stratification manifests in whether a person is in the group experiencing the transition from Industry 3.0 to Industry 4.0, or Industry 4.0 to Industry 5.0, or Industry 5.0 to Industry 6.0.

These transitions also occur against a backdrop of changing project typologies, wherein architectural form must be more adaptable in order to support people and communities, for whom the very nature of work, leisure, and activities of daily living is rapidly morphing.

THE SET OF PLACES

This book answers the fundamental question with respect to the Industry 6.0-enabled built environments in which humans work, rest, and socialize. An Industry 6.0-enabled built environment is no longer merely the passive vessel in which these technologies reside. Nor is it merely inclusive of basic, static HVAC, electrical, plumbing, fire suppression, and information technology control systems (i.e., building automation systems, or BMS). Rather, the built environment's control systems are becoming as multilayered, expansive, dynamic, and complex as the control systems of any other Industry 6.0 system of systems. These control systems include:

1. *Sensing*: sensor networks,
2. *Perception*: software systems that recognize and interpret sensed signals and prepare the data for use by the cognitive systems,
3. *Cognition*: systems that contain mental models of systems and environments and behavioral protocols, and that utilize perceived information about the

environment and the actors within it in order to monitor and classify the activities taking place within a bounded region of concern, and then determine, plan, and initiate appropriate responsive action, and

4. *Responsive Action*: a response in the form of either communication with actors in the region of concern or a physical, actuated response through the use of motors and pumps to reconfigure parts of the system to better support the activities taking place within the region of concern.

Useful points of reference, which will be expanded upon later in Part VI, are the systems used on the International Space Station. The systems include hundreds of thousands of sensors, hundreds of thousands of data points processed continually, and millions of lines of code to perceive, analyze, classify, and utilize the data and manage responsive actions. As the built environment becomes imbued with Industry 6.0 technology, this type and scale of intelligent, adaptive systems will be commonplace here on Earth as well.

THE SET OF USERS

This book answers the fundamental question with respect to a heterogeneous population of users, spanning socioeconomic strata and cultures, spread across the globe, who have regular access to Industry 4.0, Industry 5.0, or Industry 6.0 systems in their respective built environments. These users share common experiences interacting with automated systems, no matter where they live, in that the same types of computers, software, web-based systems, appliances, and other computer-based technologies mediate how they execute tasks and experience the world. The focus is primarily on the dynamics of the emerging Industry 6.0 technology set.

FACTORS INFLUENCING THE ANSWER TO THE FUNDAMENTAL QUESTION

This book addresses the fundamental question with respect to six factors, three of which are timeless, and three of which are bound to this era, its technologies, and its set of users.

1. The first (timeless) factor influencing the answer to the fundamental question is *embodied cognition*. That is, we use the world around us as cognitive scaffolding, and as extensions of our bodies and minds, to enhance execution of our physical and cognitive endeavors. This is meant in the literal sense, not the figurative sense. Andy Clark's contention is that since our species first picked up a rock to crack a nut (i.e., tool use), we have been cyborgs, and that primarily, we build, "better worlds in which to think" (Clark, 2003). While this is generally true, this book is particularly concerned with the implications of this contention in light of the emergence of Industry 6.0 technology. This book addresses ways in which the built environment serves as scaffolding for people—no matter the complexity of the environment— but also specifically, the emerging condition wherein the built environment

includes advanced, integrated control systems, AI, and robotics processing information and adapting to people and situations in real time.

2. The second (timeless) factor influencing the answer to the fundamental question is human sensitivity to tool use, including use of the built environment. Once a human uses a new tool, neuronal excitation patterns for representation of the body schema in the mind, with respect to task execution (specifically neuronal encoding of tools as extensions of one's own body and mind during task execution), begin changing to incorporate the new tool as part of its representation of the self *within minutes of using a new tool or experiencing a new environmental affordance. In many cases, the neuronal excitation patterns settle into new, mature, stable, optimized routines that include tools/environmental affordances as parts of self within days to weeks.* Given this, *we humans are highly sensitive to changes in the world around* us—especially the tools/affordances/environments we use. We treat them as extensions of our own body and mind—and this adaptation happens very quickly in response to new stimuli and new tools. *If this is true for using a rake or swinging a golf club, then what are the ramifications for using a smartphone, an interactive whiteboard wall, or reconfigurable architectural features?* This is why design matters. This is why design is an ethical act. *When people use the things that we design, we are partially responsible for reconfiguring their sense of self, in a literal way, and their perceptions, cognition, and actions—and this can be to their benefit or detriment. Therefore, we must be mindful and intentional about the impact of our designs on people.*

3. The third (timeless factor) influencing the answer to the fundamental question is that these structural and behavioral process adaptations are part of an overall species adaptation mechanism for surviving in the world, from an evolutionary perspective. More specifically, the theory of ECOLOGICAL NICHE CONSTRUCTION (ENC) (Odling-Smee et al., 2003) defines four ways that species modify themselves and their environments in order to improve their existence and survivability. This is not just true for humans. This is true for all species. This book is concerned with how the advent of Industry 6.0-based, real-time-adaptable environmental systems are utilized with respect to ENC processes. Hence, to answer the fundamental question, it is not just about the *coexistence* of certain phenomena in a particular time, in a particular set of places, for a particular set of users—rather, it is also about the *coevolution* of the elements of the system of systems (i.e., agents, their environments and their technologies).

4. The fourth (era-bound) factor influencing the answer to the fundamental question is the acknowledgment that humans experience trauma, and that there are a number of traumas that are somewhat unique to this particular time period. Some of these traumas are caused or exacerbated by deficiencies in how we deploy Industry 4.0/5.0/6.0 technology right now. We refer to these *sources of trauma*, or *sufferings*, as *colonizers* because they impose themselves upon our lives and our societies in ways that are similar

to but more pervasive than the historical concept of Western colonization. Furthermore, thoughtful design of technology can help to mitigate these traumas to a degree.

5. The fifth (era-bound) factor influencing the answer to the fundamental question is addressing the constructs, methods, and tools needed by designers in order to achieve Industry 6.0 systems development. Importantly, we'll see that achieving Industry 6.0 systems requires a shift in design strategy wherein *intelligent, adaptive* systems can only be partially designed and engineered before construction and operation. Rather, for Industry 6.0 systems development, the design must start with some designed/engineered *components* but then the overall design must be finished by *cultivating it into existence while in use*—and during this cultivation, the agents, their environments, their tools, and their social organization will *coevolve* into a fully integrated, optimized Industry 6.0 system. This is a huge paradigm shift—we *cannot* engineer Industry 6.0 systems entirely before implementing them. *It is for this reason that the book places a strong emphasis on architectural strategies that imbue built environments with the capacity for reconfigurability and multiple use options.*

6. The sixth (era-bound) factor influencing the answer to the fundamental question is about understanding that the design process evolutions that will permit us to design Industry 6.0 systems well right now are also the fundamental building blocks for designing and building in outer space at commercial and industrial scale in the near future. Thus, we must master designing for Industry 6.0 if we are ever to design for life on the moon, Mars, or long-term space travel.

RAYS OF HOPE

To answer the fundamental question, we must understand what available technologies and design constructs, methods, and tools are available right now. With respect to the factors influencing the answer to the fundamental question—and especially in response to the traumas and colonizers listed in Factor 4, we present current design strategies and technologies that can be used to mitigate the described colonizers and their associated traumas. These "defeating mechanisms" that may be used to mitigate the traumas are referred to as *rays of hope*.

THE ANSWER TO THE FUNDAMENTAL QUESTION

We shape our buildings and afterwards our buildings shape us.

(Churchill, 1941)

Humans have consistently adapted to shifting environments, technologies, and societal dynamics. So one could say, "there is nothing new under the sun"—this current adaptation is just one instance in a continuum of adaptations. This is true. And yet, the coexistence and coevolution between humans and intelligent, adaptive environments will create entirely new possibilities for how humans use and experience built environments. In the resultant, emergent paradigms, humanity and intelligent, adaptive environments will shape one another in a process of mutual transformation in much faster and more dynamic ways.

Ultimately, this book does not have a complete answer to the fundamental question. Rather, this book points to several strategies and technologies that are likely foundational to answering the fundamental question. In summary, this book posits that the answer to the fundamental question with respect to Industry 6.0 technology will require a combination of:

1. rigorous requirements engineering (RE) *and validation* processes;
2. sociotechnical systems (STS) development processes;
3. a calibrated trust assessment that places each agent along a spectrum from tool to collaborator;
4. advanced, evidence-based design (EBD) workflows that support:
   a. designing with ENC and embodied cognition in mind,
   b. a focus on trauma-mitigating and human-empowering strategies;
5. adaptive environmental programming strategies (i.e., Trans-Programming and Cross-Programming);
6. non-humanoid social robotics (NH-SR) systems;
7. Trustworthiness by both human and non-human agents;
8. Integrity by both human and non-human agents;
9. Virtue by both human and non-human agents;
10. A team of human and non-human agents within a sociotechnical ecosystem will benefit most from an allocentric, mutualistic symbiotic relationship; and,
11. Task allocation, sensing, perception, cognition, and action are enhanced by mapping and managing agent (A), agent-to-agent (A2A), agent-to-infrastructure (A2I), and agent-to-everything (A2X) relationships, within an Operational Design Domain using Operating Envelope Specifications.

The points 7 through 11 are especially important. Continuous adaptability is a messy process. All agents will make mistakes. There will be emergent situations that the human-AI-robot team of collaborators will have to work through, and the desired outcomes will not be clear without some trial and error. When they do encounter mistakes or have to go through trial-and-error experimentation to find the right next solution for an evolving situation, they have to be able to trust that the entire team are all trying to do the right thing, the right way, and that the adverse outcomes and failures are truly just mistakes and not evidence of incompetence or ulterior motives by any agents. In the context of intelligent, adaptive environments, *trustworthiness* means adhering to validated requirements without exception, whereas doing the right thing, that is, *virtue*, means always acting to achieve validated requirements and/or goals even when how to do so it is not clear (i.e., when the right course of action is not clear, there is always a strong bias toward achieving validated requirements and/or goals to the extent possible), whereas doing it the right way, that is, *integrity*, means always acting according to an established code of rules without exception. An allocentric, mutualistic symbiosis means that all collaborating agents benefit from their collaborative engagement. If trustworthiness, integrity, virtue, and an allocentric, mutualistic symbiotic bias are maintained, then trust between team members is likely to occur. And among trusted partners, mistakes can be tolerated and forgiven. But in the absence of trustworthiness, integrity, virtue, and an allocentric, mutualistic symbiosis, trust is lost, and the human-AI-robot team ceases to function effectively.

"A-DAY-IN-THE-LIFE-OF HUMANS"...2050 SCENARIOS WITH PAULETTE ROMILLY

Humans have consistently adapted to shifting environments, technologies, and societal dynamics. Today, the notion of coexistence and coevolution between humans and intelligent, adaptive environments unfolds into diverse and compelling trajectories. These intertwined pathways explore and redefine the reciprocal relationship between evolving human needs and adaptive environmental responsiveness. This is the next major step in humanity and their environments shaping one another in an ongoing process of mutual transformation.

To delve into these notions, we render the concerns more vivid by depicting alternative day-in-life scenarios for an individual in 2050. *Paulette Romilly* is a 37-year-old, renowned data analytics expert of global fintech. She primarily works from home, regularly venturing into a London corporate office once every two weeks. She journeys roughly once per month on average for work or leisure. She has generally maintained good health yet now reaches a stage where her vitality and resilience begin to wane with age, rendering her prone to long-term, chronic ailments over time.

For these reasons, Paulette must diligently care for herself to avoid a future of diminished mobility and mental acuity. Compounding her health maintenance challenges are the myriad social, technological, and lifestyle stressors she confronts daily, alongside the expansive network of computer-technology-enabled systems with which she must interact seamlessly, and that constantly compete for her attention and demand that she adapt to a changing technological milieu and associated routines for work, for leisure, and for daily living. In addition to the social, technological, and lifestyle stressors, a subtle yet vital stressor lurks all around her. The failure of Paulette's built environments and technologies to adapt and engage in stressor mitigation substantially adds to the stressors in her life. She is perpetually striving to elevate her quality of life, yet the very environments and technologies she employs to enhance her existence are excruciatingly ill-fitted, slow to adapt, and inflexible to her ever-evolving needs.

DAY IN LIFE SCENARIO IN 2050

TRAUMA-EXACERBATING SCENARIO

Paulette Romilly has a cadre of AI-based assistive agents.

Kawawi, Ikaoid, Uwx, Robbie, and Ranger (artificial agent friends) support Paulette in her modest life in Twickenham. Kawawi is Paulette's car's autonomous driving system (ADS) (and the car itself). Ikaoid is Paulette's home automation system (and the home itself). Uwx is Paulette's work and education optimization coach. Robbie is Paulette's health and wellness coach. Ranger is

Paulette's companion who aids in organizing and sharing social experiences with her other human and non-human acquaintances. For example, Ranger accompanies Paulette when she ventures out at night, in adverse weather, or attends rugby matches at Twickenham Stadium.

   Paulette faces myriad challenges in her daily life.

   In 2050, Paulette's existence entails navigating and mitigating a plethora of sufferings, including the Hoplitus virus global pandemic, wealth depreciation due to accelerating population decline, high inflation from runaway debt increases, social unrest, cyberbullying, identity theft, chronic loneliness, chronic anxiety, overwork, a scarcity of free time, and limited access to healthy diversions.

   Paulette DOES NOT TRUST that her agent allies are genuinely committed to helping her attain authentic, meaningful independence, peace, comfort, and health. She also does not understand (conceptually) how they operate, which hinders her ability to collaborate with them in fostering empowering behaviors in her daily routines. She also does not have the option to decline using the agents, since declining their use would prevent her from being upwardly mobile and socially engaged.

   Paulette's artificial friends are competitive, each originating from a different multinational corporation or government agency, and each primarily driven to collect and monetize her data (and her trauma), to surveil, to constrain her information and channel her behaviors, and to boost product sales and sway political participation with limited or no regard for what is in the best interests of Paulette. Paulette is always suspicious of the support provided by her agents. Is it really in her best interests? Is it what is in society's best interests? The combined stresses of these ongoing traumas and the inability to trust the computer-technology-enabled agents purportedly "improving" Paulette's quality of life result in maladaptive behaviors (coping mechanisms), heightened cortisol and adrenaline production, fragmented sleep, endocrine disturbances, and the emergence of chronic psychological and physiological disorders and lifestyle diseases.

TRAUMA-MITIGATING SCENARIO

Paulette Romilly has a cadre of AI-based assistive agents.

   Kawawi, Ikaoid, Uwx, Robbie, and Ranger (artificial friends) support Paulette in her modest townhome in Twickenham. Kawawi represents Paulette's car's ADS (and car). Ikaoid is Paulette's home automation system (and home). Uwx is Paulette's work and education optimization coach. Robbie is Paulette's health and wellness coach. Ranger is Paulette's companion who aids in organizing and sharing social experiences with her other human and non-human acquaintances. For example, Ranger accompanies Paulette when she ventures out at night, in adverse weather, or attends rugby matches at Twickenham Stadium.

   Paulette faces myriad challenges in her daily life.

In 2050, Paulette's existence entails navigating and mitigating a plethora of sufferings, including the Hoplitus virus global pandemic, wealth depreciation due to accelerating population decline, high inflation from runaway debt increases, social unrest, cyberbullying, identity theft, chronic loneliness, chronic anxiety, overwork, a scarcity of free time, and limited access to healthy diversions that persist in modernity truly. While Paulette does feel compelled to use these agents to fully participate in the workforce and social groups, she also feels the freedom to opt out of using the agents without concern that doing so will limit here professional and social opportunities.

Paulette TRUSTS that her agent allies are motivated to help her achieve real, meaningful independence, peace, comfort, and health. Paulette's trust is also enabled because she understands (conceptually) how they are doing what they are doing, which means that she can partner with them on affecting empowering behaviors in her daily routines.

Paulette's artificial friends are competitive, each originating from a different multinational corporation or government agency, but each with a mandated prime motive of maximizing <u>their human-AI-robot team's</u> success and well-being, minimizing <u>their team's</u> stressors and trauma, and also supporting the overall success of <u>their team's</u> community, as part of their human-non-human agent team. Legally, these agents are treated as utilities, the same as potable water, wastewater removal, electricity, and the internet, and managed independent of economic or political concerns. Practically, their primary focus is the optimization of Paulette's health and well-being, as well as their own, with the understanding that optimizing Paulette's human-AI-robot team is in the best interests of Paulette, her community, the economy, and the greater society. Paulette is content with the support provided by her agents. The agents adaptively support Paulette's emotional, social, and physical well-being, particularly during traumatic periods. Embedded within a sophisticated cybernetic "artifact ecology" (Kirsh, 2001) of robotics, Artificial Intelligence (AI), and Cyber-Physical Systems (CPS), the agents represent a new generation of intelligent, adaptive environments, systems, and services. These advanced technologies are seamlessly integrated into daily life, alleviating social, physical, and environmental stressors, and promoting overall wellness.

FROM TRAUMA, SUFFERINGS, AND COLONIZERS TO RAYS OF HOPE AND THE ALL-IN-ONE ENVIRONMENT

The juxtaposed "day-in-the-life-of" scenarios illustrate that the nascent practice of designing "intelligent, adaptive" environments is at a crossroads. Will it develop to provide centralized, provider-centered supervisory control, surveillance, and exploitation? Or will it develop to provide decentralized, user-centered performance optimization? Complicating the response to these questions is the tendency for current "smart" environments and technologies to exhibit limited dimensionality and adaptability, with ambiguous (at best) and malevolent (at worst) motives and processes. Given these points, the way we design, develop, and employ intelligent, adaptive environments and technologies stands on the brink of a significant evolutionary leap. From environments (e.g., restaurants, museums, shopping malls, homes) to technologies (e.g., XR, "smart" appliances, ADS, Instagram, Facebook, X [formerly Twitter], ChatGPT), existence now encompasses both extraordinary new opportunities for personal and social engagement and potentially dreadful personal and social stressors and traumas.

It is due to these concerns that the "coexistence and coevolution" thesis articulates and contemplates current advancements in architecture with respect to parallel developments in robotics, human factors, cognitive science, systems engineering, AI, and technology, employing literature reviews and case studies to uncover strategies for alleviating modern social sufferings through Industry 6.0-enabled intelligent, adaptive environments. This book's content is a work in progress—a theoretical and experimental exploration into these subjects—designed to formulate pragmatic solutions to intricate social and environmental challenges intensified by today's rapidly evolving technological landscape. What follows are detailed portrayals of the principal human "sufferings" at stake, the technological and methodological "heroes" poised to mitigate these afflictions, and a proposal on harnessing these heroes to alleviate such sufferings.

The six parts of this book elaborate on these concerns and present how they align with the emerging methodological and technological trends of trauma-informed design (TID), embodied cognition, STS development, human-centered design, and Industry 6.0 technologies. The content is as follows:

Part I: Colonizations at the Age of Transformations
Part II: Heroes of the Past and the Present
Part III: Sufferings of Humans in the 21st Century
Part IV: Rays of Hope/Heroes of the Near Future
Part V: Designing Coexistence/Coevolution
Part VI: Epilogue: All-In-One (AIO) Environments

PART I: COLONIZATIONS AT THE AGE OF TRANSFORMATIONS

The present and near-future worlds are imperiled by a new form of colonization. The colonized "sufferers" are humans and their personal and group activities and relationships. The colonizers traumatizing the sufferers are social, environmental, and health crises as mediated and exacerbated by an ever-changing, fragmented collection of technologies. Part I (Colonizations at the Age of Transformations) describes how crises beget traumas, including social (e.g., loneliness, anxiety, fatigue, inequality), environmental (e.g., socioeconomic, educational, and resource disparities, as well as adverse impacts on the biosphere), and health-related (e.g., lack of physical activity, lack of good quality food, poverty, diseases, and pandemics).

PART II: HEROES OF THE PAST AND THE PRESENT

Throughout human existence, humans have conceived myriad systems to overcome traumatic challenges, spanning social, environmental, and health domains. These contemporary "heroes" bolster humanity in mitigating these challenges. Part II, "Heroes of the Past and the Present," delineates how these interventions ensure the survival of the species, but can also be traumatic for individuals, particularly by neglecting human variability and complexity—and in so doing, generate additional stress and trauma upon users.

Part III: Sufferings of Humans in the 21st Century

Part III (Sufferings of Humans in the 21st Century) elaborates on present external crises and the concomitant traumas spawned by our environmental and technological remedies. The limitations of our current technological "heroes" (e.g., Industry 4.0 technologies) lie in their compounded complexity, fragility, and an inherent technology-centric disposition; their dearth of adaptability; their failure to achieve a true goodness-of-fit to the genuine nature of the underlying needs they address—often emerging from developers' inadequate mental models—and their erratic, ill-fit developmental processes. The fragmented and uncoordinated evolution of these environmental and technological interventions, together with the concomitant expectation that humans should compensate for their deficiencies, exacerbates human suffering.

Part IV: Rays of Hope/Heroes of the Near Future

Part IV (Rays of Hope/Heroes of the Near Future) investigates approaches to assuage the immediate impacts of crises and to do so in manners that forestall inadvertent traumas inflicted by the environmental and technological solutions themselves. The advent of Industry 4.0/5.0/6.0 technologies heralds the birth of adaptive, resilient environments and innovations, emerging through both nascent and maturing human-centered methodologies. "Rays of Hope/Heroes of the Near Future" outlines several nascent methods that are useful for designing Industry 6.0 systems.

Part V: Designing Coexistence/Coevolution

Part V (Designing Coexistence/Coevolution) presents a vision of human/non-human agent/environment symbiosis, as well as case study projects that provide insights into designing intelligent, adaptive environments. Through the "Designing Coexistence/ Coevolution" case studies, the "Heroes of the Near Future" are brought to life in practical application.

Part VI: Epilogue: All-In-One (AIO) Environments

Part VI: Epilogue: All-In-One (AIO) Environments presents hypothetical scenarios wherein every challenge confronting humanity is addressed by integrated teams of human–environment–technology collaborators. Paulette Romilly, our heroine, contends with a series of crises. In the AIO Environment, Paulette safeguards herself from external threats by co-adapting through choreographed behavioral routines that embody deeply integrated and symbiotic interactions among Paulette, her environments, and her technological agents. Their relationships are inextricably interlinked, fluid, flexible, and responsive. The AIO system of systems mitigates a multitude of traumas by rapidly adapting and establishing new goodness-of-fit models between the human/environment/technology ecosystem—the *intelligent artifact ecologies*— and the external challenges it faces.

Part I | Colonizations at the Age of Transformations

COLONIZATION...

We should speak about colonization openly, honestly, and gently. The term itself is heavy, evoking images of ships coming uninvited, flags claiming land already home to other voices, and outsider's languages and laws forced upon people yearning to preserve their own stories and governance. However, if we believe colonization ended with the fall of empires, we are only seeing part of the truth. There were subtler aspects to colonization than just the use of raw, physical power for imposing the colonizers' frameworks of rules. There were incentives and meritocracies and upward mobility, used to entice the colonized to be complicit in their own subjugation, to self-police, and to self-enforce the colonizer's systems—all for the prospect of earning a bit of freedom or privilege. From this perspective, many of these "reward-based" vestiges of colonization persist today, adapted for use influencing constituencies and customers. Colonization didn't vanish; it transformed. While its methods have changed, its quiet and pervasive influence continues to exist.

Historically, colonization involved claiming land and territories and subjugating peoples with force and with laws. Nowadays, colonization is more about influencing our thoughts, feelings, actions, and interactions with the world as mediated through social media technologies and access to ecosystems of information, products, and

 DOI: 10.1201/9781003441953-2

services. The colonizers today don't wear uniforms or fly flags; they operate within the subtle mechanics of code, digital platforms, algorithms, supply chains, licensing systems, surveillance systems, access to telecommunications and information networks, and control over systems that observe, predict, and guide our behaviors. Rather than a single empire imposing its will, we now face a complex web of influences from major corporations, government agencies, criminal enterprises, and nongovernmental organizations, all shaping our lives in ways we often overlook because we feel powerless to change the dynamics.

This isn't merely a figurative description; it's our reality.

Even dictionaries associate colonization with the past, linking it to distant places and direct foreign rule. But what occurs when colonization doesn't rely on a physical presence? What if our surroundings quietly collect data, our routines follow unseen scripts, and algorithms gently steer our choices without our explicit consent? Or worse, what if all of this happens *with* our explicit consent because we had to first grant consent to be observed and influenced before we were allowed to access the technologies or networks required for upward mobility and socializing? This new form of colonization doesn't loudly announce itself; it seamlessly integrates into our daily lives, subtly shaping our worlds from the background.

We refer to this concept as *Poly-Colonization*—a term that captures today's subtle systems of surveillance and control—and in response, we urge greater awareness. The struggles of the past persist today through sensors, screens, and influence campaigns and news cycles that quietly constrain our gestures, emotions, and decisions to align with unknown entities' agendas. On a global scale, telecommunications technologies and social media have taken on a very large role in shaping our identities, relationships, and sense of reality.

And yet, as the saying goes, there is nothing new under the sun. Technology, including architecture, has always been co-opted to project power, to control, and to influence. Tools, infrastructure, and architecture don't just support us; they subtly orchestrate our movements, perceptions, experiences, and senses of self. The difference now is the rapid rate of change of the messaging and the malleability of the media (even real-time change). Architecture, once defined by solid materials and applied, static finishes, now intertwines with these subtle and dynamic influences. Buildings have evolved to integrate sensors, algorithms, surveillance mechanisms, and a continuous barrage of visual and auditory cues. We exist within a complex ecology of control that is locally dispersed and globally centralized, incessantly suggested rather than imposed. Instead of a dominant culture overtly enforcing its will, we experience multiple systems of influence quietly nudging our behavior, providing choices that seem self-directed but are in fact constrained by what is beneficial to unseen influencers, thereby, subtly restricting our outcomes. Our complicity is key to the illusion that these are *our* perceptions, *our* choices, and *our* consequences.

Contemporary colonizers wear friendly faces: convenience, personalization, self-empowerment, optimization, and security. *We invite them* into our homes and communities, forming bonds without fully understanding their implications. Yet, this closeness makes critical reflection even more essential—and difficult. Colonization flourishes when it feels natural, inevitable, and unobtrusive—"same as it ever was" (Byrne et al., 1980).

There is no such thing as neutral design. Every interface affects us; each algorithm subtly guides our desires and behaviors. Even the most thoughtfully designed "smart city" can reflect hidden power dynamics and control. The data harvested from our lives and the predictions regarding our actions raise serious questions about the very nature of freedom in our modern world.

Heidegger warned us about losing touch with authentic human experiences as we increasingly rely on technology. Eliot cautioned, "Where is the wisdom we have lost in knowledge? Where is the knowledge we have lost in information?". And as computer scientists add, "Where is the information lost in data?" As our surroundings become more actively and dynamically responsive via integrated sensors, processes, and actuators, the authenticity of space—its history, memories, relationships, the trustworthiness of our experiences, and our capacity for reflection—is often overshadowed by others' vested interests in our thoughts and behaviors.

Decolonizing today goes beyond removing overt control; it involves interrogating systems that seem normal, innocuous, or even helpful. It means reclaiming the messy intricacies of human life and social relationships, and protecting our right to be unpredictable, complex, and even inefficient. These seemingly "inefficient," "messy" behaviors are part of humans developing creativity, flexibility, resilience, stamina, and mastery over skills, rules, and knowledge.

Bernard Tschumi characterized architecture as spaces filled with possibilities, rather than just physical structures. Deleuze and Guattari proposed the rhizome—a fluid, networked, non-hierarchical model—as a means to resist dominant frameworks. Jacobs championed the thoughtful use of environmental design features to create and strengthen communities. Imagine designing spaces based on these principles: open-ended, adaptable, human-scaled, nurturing salubrious psychological and social outcomes, defying rigid control.

We must recognize and cherish the overlooked spaces—the quiet corners, informal paths, improvised shelters, and subtle human interactions often neglected by formal design. These aren't failures or signs of neglect or deterioration; rather, they're courageous acts of autoregulation and creativity. We must cherish and cultivate places where people affirm their humanity despite systemic pressures.

Reframing colonization in our time means rejecting its subtle forms. It involves asking questions to understand what it means to design without exerting dominance; to occupy space with respect, humility, empathy; and to engage technology in a measured, purposeful way, without being controlled by it. This reform is vital because today's colonization often remains invisible, operating quietly as infrastructure. Our responsibility is to illuminate what has become obscured in order to create futures that embrace diverse ways of being instead of merely efficient systems as defined according to the vested interests of only a select few.

In essence, contemporary colonization—and overcoming it—is about the agency to moderate our own perceptions and shape our own lives. If architecture and design are to avoid becoming subdued instruments of dominance, they must evolve into spaces fostering genuine user-directed, unmediated openness, collaboration, and creativity. The future demands not only new designs but also fresh ways of thinking. It begins with questioning not how we occupy spaces, but how we might liberate them.

IN TRANSFORMATION...

POLY-COLONIZATION: AN EXPANDED DEFINITION

Poly-Colonization encompasses a confluence of forces—global health challenges, viral mutations, advanced gene editing techniques, increasingly complex virtual platforms, and hidden layers of authority, all leveraged by unknown interests in a sociopolitical milieu of extreme inequality—that alter personal and collective behaviors in subtle yet profound ways.

Poly-Colonization is a mesh of distinct but overlapping layers: systems, agents, needs, wants, resources, debts, territories, and perspectives. We continue to believe we operate by free will, yet the scaffolding of modern systems steers our preferences, behaviors, and social ties. Scholarly voices emphasize that this subtle orchestration can be as powerful as outright force. Shoshana Zuboff (2019) highlights "surveillance capitalism," where personal data is monetized to shape choices, while older theories of power from Michel Foucault echo in the background, illustrating how life itself can be directed and molded.

As Industry 4.0/5.0/6.0 technologies become indispensable, our reliance on them prompts deeper questions about dependency, trust, and the future of human agency. We stand at a threshold where the merging of biological, technological, and cultural forces redefines what it means to be free, connected, and in control.

PAULETTE'S COLONIZERS

Paulette began her day like any other—her alarm synchronized with her sleep cycle, her window blinds timed to open precisely as the sun rose. She'd always found comfort in the hum of seamless automation, from the fridge that preemptively ordered milk to the exercise app that monitored her vital signs. Yet beneath that aura of convenience lay an undercurrent of unease. Every device she cherished was also a portal into her personal world, compiling data that fed into unseen repositories, moderated and leveraged by unseen economic and political powers. Though she marveled at her time-saving routines, she often caught herself wondering: "Am I shaping these systems, or are they shaping me?"

Her life pivoted dramatically on a day that should have been routine. She woke to find her data-assistant console flashing an urgent prompt: a new city ordinance mandated upgraded biometric scans for public transit. She felt an unexpected twinge of dread. When she walked outside, scanning her wrist to board the bus, a slight error triggered more questions. A swarm of lights blinked on the display, culminating in a halting of service. The system did not recognize her as the owner of the bank account that she was using to pay for bus fare, and so it refused to process the transaction. She had become a data anomaly.

Instantly, her neighbors' curious glances multiplied. Gone was the usual mild chatter about weather or local events—her glitch made them uneasy, as though she harbored some intangible flaw in a world that demanded seamless data flows. She could see in their faces that they were all wondering, "What is she

hiding? What did she do?" Frustrated, she trudged down the street alone, notic-
ing how the city's towering screens now assaulted her with generic ads—not the
personalized ads to which she had become accustomed.

Desperate for solace, Paulette retreated to her home, only to find more
disquiet. Her smart thermostat refused to override the factory-recommended
temperature settings because it did not recognize her. Her AI-driven home
automation did not recognize her, and so classified her as an intruder, and
demanded multiple voice, biometric, and key codes for her to verify her iden-
tity or else it would lock her in and call the police. Her health and wellness
assistant, not recognizing her, attempted to onboard her as a new client. In
the background, news bulletins scrolled across her screen, touting the city's
success in maintaining order through predictive analytics and advanced
infrastructure. She realized that the same apparatus that once cradled her
in efficiency now served as an omnipresent gatekeeper—silently guiding her
health, finances, and even her daily mood.

A friend recommended she seek assistance at a specialized "Center for
Personal Autonomy," a municipal outgrowth designed to help citizens navi-
gate the complexities of automated systems. As she stepped inside, Paulette
felt a jarring mixture of relief and disorientation. Rows of tablets lined the
walls, each offering a unique "solution" to algorithmic oversight. Counselors,
themselves reliant on digital tools, asked her to input every detail of her day
so they could identify where the system "misread" her data. Embarrassed, she
consented, unveiling private habits.

When she emerged, the bus pass glitch was resolved, but Paulette felt less
triumphant than ever. She understood that her city's polished efficiency was
both a wonder and a veiled clamp on agency. From that moment on, she
vowed to tread carefully, weaving her life through streams of data that, at any
instant, could tilt her world into new forms of dependency and restriction.
Behind every convenience lurked the tension between progress and surrender,
a reminder that, in this new era of Poly-Colonization, each step forward might
carry a hidden cost.

PHILOSOPHICAL PERSPECTIVES ON POLY-COLONIZATION

Understanding Poly-Colonization calls for an engagement with thinkers who scruti-
nize the delicate interplay between technology's promise and its potential for covert
domination. Michel Foucault remains pivotal. His exploration of biopower—where
control is enacted not solely by force but by regulating human life itself—resonates
in a world permeated by algorithms. Instead of open coercion, modern networks
steer conduct through "nudges" disguised as personal choices. Foucault's Panopticon
metaphor (1977) now echoes in digital platforms that collect and interpret granular
data on every individual. Though we might now remain physically unrestrained, the
knowledge that we are observed at any point encourages self-policing. De Cauter
(2004) cautions that our surveillance and media technologies prevent us from true

interpersonal and civic engagement. Arendt (1958) cautions that we have devolved in a social realm of compelled behaviors born of necessity and escapism, without the agency to act independently. Meanwhile, Shoshana Zuboff (2019) advances the notion of surveillance capitalism, an economic order where human experience is treated as raw material to be commodified.

Gilles Deleuze (1992) built on Foucault's foundation, describing "societies of control" in which regulation is fluid and continuous, not enclosed within static institutions. In a digitally transformed environment, protocols shift in real time, adapting to user profiles and contexts. The result is a decentralization of power, but one that is, in many ways, more pervasive. No single force brandishes control; instead, an entire network of devices, server farms, and data gateways orchestrates how we spend, learn, or interact. The result is a reimagined version of Foucault's biopower: those with access to immense data can shape not just what we purchase, but also our social and political allegiances. If we pause to consider how quickly we consent to data collection for convenience or entertainment, we see the subtlety of modern colonization. The question is: Are we users of technology, or products in someone else's grand marketplace?

Amid these reflections, it becomes evident that the nature of power has shifted to intangible, digital networks. Indeed, Poly-Colonization roots itself in precisely these intangible oversights, where massive platforms appear to "serve," yet systematically steer. The user's sense of "choice" persists, but any real friction lies concealed in code, beyond typical awareness or comprehension.

...PARASURAMAN: NEUROERGONOMICS

Central to these transformations is the tension between augmenting human capabilities and eroding human agency. Parasuraman's research on neuroergonomics (2000) zeroes in on how the brain interacts with computational aids (Parasuraman & Rizzo, 2007). At first glance, automation is benign—improving efficiency and minimizing error in tasks like flying a plane or handling complex data. Yet, as Parasuraman explains, over-reliance on automation can degrade situational awareness, human mastery of skills and knowledge, and also stifle adaptability and resilience. Parasuraman proposes instead to develop systems for "adaptive automation," wherein they *augment* human task performance only to the minimum extent needed to maintain a targeted level of performance based on real-time assessments of user workload, capability, performance, or stress.

Jacques Ellul (1964), writing decades earlier, warns that as technology pursues absolute efficiency, society's creative spirit and freedom can be undermined. By linking Ellul's broader critique to Parasuraman's narrower focus on automation's impact on cognition, one sees how being "helped" by machines slowly morphs into dependence. Over time, people lose the competence—or will—to override a system's recommendations. Neuroergonomics thus becomes a frontier that underscores both the power and the peril in forging alliances with advanced tools. A well-designed system can amplify human potential; a poorly designed one can colonize, stultify, and/or diminish task performance, adaptability, and resilience. For these reasons, the systems with which we interact have the power to enhance or undermine our perception of our own agency and identity.

...On Trust and Distrust

Trust in technology is a precarious bargain. While the uniform precision of algorithms can tempt us to defer to machine judgment, the real world regularly exposes the pitfalls. Military studies have revealed soldiers ignoring their instincts in favor of robotic scanners that declared certain zones "safe." In some instances, that confidence led to real danger. Healthcare, too, has had incidents where automated systems for prescriptions or diagnostics introduced bias or errors, all while medical professionals—trusting "objective" software—rubber-stamped flawed recommendations.

Balancing trust and skepticism is vital for ensuring that people can intervene when algorithms malfunction. Over-reliance on software means diminishing our own capacity for situational judgment and adaptation. Yet, total distrust stalls progress, rendering potentially life-saving or efficiency-boosting tools obsolete. Poly-Colonization's cunning lies in how swiftly it normalizes reliance on digital processes. By the time we question whether to trust or doubt, we have already assimilated the machine's presence into daily life.

...On Language

Language is never a neutral vessel; it carries historical and cultural freight that molds our perceptions of what is "modern" or "backward." Poly-Colonization operates not merely by controlling physical resources but by influencing how people describe and interpret technological expansion. Terms like "development," "opportunity," or "opening" may sound benign, but they can smuggle values and power relations that favor certain groups' interests.

In Arabic, for instance, isti'mār (colonization) often conjures memories of external domination, cultural devastation, and financial exploitation. However, advanced technology might be discussed within the scope of tanwīr (enlightenment) and may sometimes be discussed in the form of fath (opening), bringing more layered meanings. Tanwīr signals a break from the darkness of ignorance, stepping into the glow of knowledge—an uplifting notion of becoming wiser and more adept. Likewise, fath evokes new frontiers, the unlocking of uncharted potentials. Both suggest a broadening of horizons, yet they carry tensions. The "light" of tanwīr can blind as much as it illuminates, imposing a contemporary vision shaped by Western ideals, which may misalign with local traditions. Meanwhile, fath can entail a rupture from one's heritage—simultaneously liberating and wounding. Technology's "opening" to global flows may bring cultural erosion alongside fresh possibilities.

Nor is this ambiguity about how to understand the imposition of new ideas, behaviors, and systems exclusive to Arabic. Japanese, for example, uses 開国 (kaikoku) for "opening the country," recalling the forced Western intrusion of the mid-1800s that catalyzed rapid modernizing but at the cost of dislodging time-worn customs.

Chinese has 开明 (kaiming), or "enlightenment," carrying hints of borrowed foreign values that can weaken age-old Chinese frameworks. French, through "lumières," references the Enlightenment but also reminds us of how "civilizing missions" rationalized colonizing distant lands in the name of progress. Gayatri Chakravorty Spivak's "epistemic violence" describes how substituting local knowledge with a universal

Western blueprint can effectively silence or diminish native learning. Edward Said's writings on Orientalism further clarify how language can turn entire populations into exotic Others in dire need of modernization by outsiders. These attitudes survive in words like "development," "modernization," or "opening," insinuating that "non-Western" cultures must reshape themselves under a Western lens.

Edward Said (1978) and Gayatri Chakravorty Spivak (1988) illustrate how discursive "openings" often entrench the perspective of the more powerful party, diminishing local knowledge systems. Thus, language can become a subtle colonizer in its own right, reinforcing the narrative that certain technologies represent unimpeachable progress. Where older colonialism was enforced with the sword, modern variations slip through the intangible realm of discourse, embedding themselves in hearts and minds. Given the above, and with respect to contemporary "digital" colonizers, how are we to interpret their impositions on our lives? Enlightenment? Exploitation? An ethically more complex, nuanced situation, that is simultaneously both beneficial and detrimental? And how do we take control of these processes and tune them to society's and people's best interests?

...Decentralized Nature of Power

When we speak of Poly-Colonization, we must confront how contemporary power disperses across many nodes. Conventional colonization featured a visible hierarchy—a colonial center commanding distant territories. Economic imperialism harnessed monetary dependencies, while cultural hegemony manipulated norms and ideals. In contrast, Poly-Colonization thrives in the background, weaving itself into social media feeds, public infrastructure, and AI-driven systems. The "center" is no longer a single empire or corporation but an array of data flows and algorithmic logics controlled by unknown others.

This decentralization makes resistance elusive. Where does one protest if control arises from an amorphous network of corporate interests, government mandates, influencers, gatekeepers, and automated protocols? Individuals who might have confronted an oppressive regime now face intangible codes that direct millions of micro-interactions. With each new dependency—whether it is a phone that monitors biometrics or a service that auto-renews subscriptions—we become entwined with invisible structures that transcend borders.

Poly-Colonization thus exemplifies the ultimate paradox: the more intangible the authority, the more entrenched and pervasive it can become.

...On Identity and Agency

A critical dimension of Poly-Colonization concerns the subtle reconfiguration of personal identity. Social media platforms increasingly serve as our public square, but what we encounter online is pre-filtered by algorithms that "learn" our preferences and reflect them back to us in order to influence us to think and act in certain ways. Rather than broadening our worldview, these personalized feeds tend to confirm our established beliefs. The information presented to us is akin to the distortions in funhouse mirror reflections. But whereas experiencing real funhouse mirrors is an occasional, brief, diversion, providing only a momentary break from reality, these new, digital "funhouse mirrors" are pervasive and inescapable, all of the time, making the distortions seem to be the reality. Over time, we internalize these curated perspectives, mistaking them for organic self-discovery. Byung-Chul Han (2017) alludes to digital-driven "psychopolitics," where psychologically targeted stimuli narrow the individual's sense of possibility.

Agency—the sense of being the genuine author of one's actions—faces an even more fraught challenge. Advanced systems aiming to "assist" inevitably shape behavior. From streaming services that anticipate our taste in movies to job platforms that steer us toward specific roles, these conveniences nudge us down well-worn paths. Zygmunt Bauman's concept of "liquid modernity" (2000) suggests that while modern life promises choice, it also dissolves stable reference points, pushing us to accept the pre-coded templates offered by technology. Even the simplest daily actions—choosing a restaurant or a romantic partner—are guided by star ratings or recommendation engines. Over time, one might wonder: Do we still forge our own paths, or are we stepping in footprints laid out by an invisible presence with ulterior, self-serving motives?

...In Art

Art frequently anticipates social transformations. Confronting Poly-Colonization, many contemporary artists reveal how technology, once seen as liberating, can function as a mechanism of quiet subjugation. Trevor Paglen's photographs and installations expose the infrastructure of surveillance and data capture—unmarked buildings, satellites, fiber-optic lines—that facilitate a culture of watchfulness. His oeuvre implies that the natural landscape is no longer truly "natural" once it's laced with hidden eyes.

Hito Steyerl critiques the contradictory nature of visibility in digital culture, dramatizing how the quest for "transparency" can itself become a form of constraint. Rafael Lozano-Hemmer's interactive projects highlight how technology can foster connection across divided spaces, yet also underscores that those same systems can codify or deepen social fault lines. Ai Weiwei, meanwhile, challenges political censorship using global digital platforms, illustrating the double-edged sword of connectivity: it holds the power to galvanize protest, but also to surveil entire populations.

Within these creative expressions lies an unmistakable tension: technology can be a canvas for collective empowerment or a Trojan horse of hidden controls. Art becomes a sphere to question which path we are treading and whether we can reclaim interpretive power over the systems we inhabit.

...On Trauma and the Self

The psychological toll of Poly-Colonization becomes evident when individuals grapple with an unrelenting sense that their most intimate spaces—thoughts, private communications, everyday routines—are under constant observation or subject to relentless external shaping—thereby exploiting them and subsuming their agency and defining their identity. This infiltration can induce persistent anxiety. Over time, the conflation of self with system introduces a subtle estrangement: as we adapt to digital norms, we simultaneously lose touch with organic, personal rhythms.

Communities are not spared this turmoil. Rapid shifts to AI-mediated communication risk weakening traditional social support networks. Physical gathering spots—from neighborhood cafés to public parks—often yield to digital enclaves, lacking tactile warmth and communal engagement. Uneven access to these platforms further compounds the trauma for those excluded, intensifying inequalities. The result can be a deep fragmentation, where subcultures revolve around proprietary apps, each group colonized by its chosen "master technology," seldom overlapping in meaningful ways.

...On Architecture and the Built Environment

Architecture under Poly-Colonization no longer confines itself to bricks and mortar; it increasingly fuses with sensory technology, data analytics, and shifting user inputs. Malcolm McCullough (2005) describes how the digital realm merges with public space, turning everyday surfaces into interactive zones. Buildings and streets become conduits of data—through sensors, cameras, and integrated interfaces—that track movements and gather intelligence about dwellers.

Carlo Ratti (2016) proposes the concept of "open-source architecture," envisioning spaces in perpetual beta, modifiable by their inhabitants. While such adaptability fosters community co-creation, it can also generate a new layer of oversight if each adjustment is logged and analyzed by remote servers. Foucault's claim that architecture wields power by regulating human movement finds renewed significance here. Given this condition, Lieven De Cauter's idea of a "capsular civilization" (2004) describes how society retreats into private pods, often facilitated by digital convenience. People may feel safe, but the intangible walls also limit genuine human contact, fracturing civic cohesion. Poly-Colonization thus challenges architects to think beyond mere form. They must weigh how an environment's embedded technology shapes interactions, fosters or inhibits communal ties, and redefines an occupant's sense of control and of self. Where modernization once meant building taller or sleeker, it now involves carefully balancing the synergy between technological empowerment and the safeguarding of human agency.

In sum, each domain—philosophy, neuroscience, trust, language, power, identity, art, trauma, and architecture—reveals how Poly-Colonization seeps into our routines. Its methods are not as blatant as occupying forces of the past, yet the cumulative result can be just as potent. By analyzing its variegated forms, we gain the opportunity to consciously navigate the labyrinth of modern progress, potentially reclaiming the agency that often slips beyond our grasp in the digital age.

AT THE AGE OF TRANSFORMATION...

In this transformative moment, the idea of Poly-Colonization offers a way to see how numerous forces mold, manipulate, and reframe human life. This section reviews five primary colonizers in the modern landscape: pandemics and viruses, advances in genetics, digital and cyber control, the reshaping of health, and environmental and ecological effects.

COLONIZER I: PANDEMICS AND VIRUSES

Pandemics and viruses redefine how we perceive power and space. Rather than conquering by military might, these biological agents quietly alter everyday life, allowing developers and regulators to rapidly reshape public policy, social rituals, and norms, with little to no oversight. Pandemics, viruses, and resultant policy and social changes reshape architecture and urban planning best practices. Historically, illnesses like the Black Death or Spanish Flu reshaped urban layouts, but COVID-19, SARS, and other outbreaks underscore how quickly physical and digital realms entwine to enforce new rules (World Health Organization, 2020). Emergencies become catalysts for expanded surveillance, changing once-familiar streets into zones of caution and regulation (Agamben, 2005). Buildings intended for congregation morph into sealed environments, and ordinary homes take on hybrid roles—office, schoolroom, and medical station. Where past colonizers flew flags, pandemics bring mandates and digital contact-tracing. They provide a convenient excuse for power brokers to momentarily eschew rules, regulations, and rights. Pandemics allow for rapid reconfiguration of economies, regulations, civic order, rights, and the collective psyche. In this landscape, viruses emerge as potent orchestrators of modern life.

The overlay of misinformation adds another dimension to pandemic-driven colonization. Manufactured distrust in scientific authorities can polarize communities, with public squares transforming into sites of protest or conflict. Where once these spaces symbolized collective gathering, they can now fracture along ideological lines. Designers and planners thus confront not only physical hazards but also intangible social fractures, where architecture becomes a battleground for conflicting beliefs about safety, science, and freedom of movement (McNeill, 2020).

Historically, pandemics have always shaped architecture. The 14th-century Black Death spurred early quarantine practices like Venice's Lazzaretto Vecchio (Gensini et al., 2004), embedding specialized isolation buildings into urban layouts. The Spanish Flu of 1918 later influenced modernist preoccupations with hygiene and light, culminating in projects such as Alvar Aalto's Paimio Sanatorium, famous for its patient-centered design (Colomina, 2019). While these measures advanced healthier spaces, they often benefited privileged groups, leaving broader socioeconomic rifts unresolved. COVID-19 continues this legacy, prompting designs that integrate ventilation strategies, social distancing protocols, and remote-working facilities into everyday architecture. Office buildings incorporate flexible layouts that accommodate hybrid in-person and online work, while transport hubs reimagine passenger flow, with reduced crowding and improved air filtering (International Energy Agency, 2021).

Agamben (2005) warns that normalizing emergency powers risks perpetuating a culture of pervasive oversight. Rather than constructing inclusive, communal environments, societies risk reinforcing isolation, with each new viral surge tightening control measures. The challenge for architects, therefore, is to cultivate designs that champion public connection without sacrificing responsiveness to emergent health threats.

Paulette once found comfort in her city's gentle rhythms: bustling cafés, bright office lobbies, and lively sidewalks. Then came the virus, slipping through borders unseen. A sudden government announcement declared lockdown; in moments, Paulette's freedoms vanished. Her vibrant routine dissolved into stark isolation, where neighbors who once smiled warmly now crossed streets to avoid contact—their only possible interrelationships mediated through web portals and smart phones.

COLONIZER II: GENETIC DEVELOPMENTS

Colonizer II, rooted in genetic developments, represents a potent yet discreet force within the Poly-Colonization landscape—reshaping societies, ecosystems, and architectural spaces at the microscopic level. CRISPR-Cas9, Genetically Modified Organisms (GMOs), and precision gene therapies entwine scientific progress and human biology with commercial imperatives, dictating how communities interact and infrastructure evolves. As Sheila Jasanoff (2016) suggests, regulating such breakthroughs demands more than technical expertise—it requires grappling with cultural, ethical, and political ramifications. These shifts underscore how genetic interventions quietly exert control beyond overt military or economic ventures, refashioning both human bodies and the environments they occupy. From specialized labs to genetically influenced crops, the architectural responses to genetic colonization reveal a tapestry of profound social inequities and ethical quandaries.

Yet this phenomenon transcends medical complexes. Residential areas may adopt stricter zoning regulations for biotech expansions, altering once-mixed neighborhoods into sterile enclaves near advanced labs. Rural areas demonstrate another face of genetic colonization. Gene-edited crops might promise higher yields, but the resultant landscapes—immense, homogeneous fields—diminish biodiversity and cultural heritage. Local farmers risk financial dependence on patented seed cycles, intensifying economic disparities (Berg et al., 1975). Such transformations reverberate through supply chains that feed urban supermarkets, revealing how the genetic realm weaves into everyday routines. Within workplaces, gene-focused expansions reconfigure spatial usage. Tech campuses constructing biotech labs merge office pods with sterile zones, bridging conventional corporate environments and facilities designed to BSL-3 or BSL-4 standards.

Altogether, these strands demonstrate how genetic developments transcend labs and farmland, permeating everyday architecture and lived experiences. Confronting this colonization calls for architects and planners to evaluate the wide-ranging implications of gene-centric transformations, from strict biosafety codes to inclusive public dialogue spaces. Thus, whether shaping top-tier research complexes or rural co-ops, the architecture of a genetically influenced world must uphold human dignity, ecological integrity, and equitable resource distribution in the face of monumental scientific power.

Paulette cherished summer mornings in her neighborhood's communal garden, where families gathered among lively flowers and budding vegetables. Then came the biotech startup, installing an unassuming research wing at the edge of the property. Locals heard whispers of experimental gene-edited seeds, touted to grow faster and withstand harsher climates. Enthusiasm turned to unease when Paulette's young son developed unexplained rashes after playing near newly planted, neon-hued sprouts. Signs were posted overnight: "Restricted Access—Authorized Personnel Only."

COLONIZER III: DIGITAL AND CYBER COLONIZATION

Colonizer III underscores how digital and cyber forces have infiltrated even the most intimate corners of everyday life. Far from mere conveniences, smartphones, wearable devices, and immersive platforms now mediate how individuals relax, perform tasks, and interact socially in what once were purely physical spaces. People gather in living rooms yet remain simultaneously absorbed by multiple digital feeds, exchanging messages with others mere steps away. The emergence of augmented reality headsets exemplifies the drift toward an environment where physical boundaries blur into virtual overlays, making face-to-face encounters precariously optional. When digital layers wrap around daily routines, casual conversation can devolve into a silent chorus of notifications. This subtle infiltration echoes broader "Poly-Colonization," transforming not only cityscapes but also local haunts like cafés, restaurants, and community centers. Ultimately, Colonizer III compels reflection on whether technology fosters connectivity or simply crowds out shared experiences, leaving individuals physically together but socially and emotionally partitioned by personalized interfaces.

As Sherry Turkle (2011) argues, technology can create the illusion of connection while diminishing authentic encounters—an observation now visible at dinner tables where family members occupy separate digital worlds. Even cozy living rooms can become silent auditoriums, with occupants glued to devices, occasionally glancing at one another but rarely engaging in sustained conversation. Such behaviors echo the broader concerns raised by Shoshana Zuboff (2019), who warns that the commodification of digital experiences recasts everyday life into data streams, subtly shaping personal routines around corporate profit motives. Erving Goffman's (1959) notion of performance resonates in these scenarios: what used to be social interplay can become an endless loop of curated digital "fronts," performed for onlookers both online and off. The more advanced these devices become, the greater the temptation to prioritize curated digital interactions at the expense of tangible, unmediated human contact.

Cafés, bars, and restaurants—once central to forging local identity—offer striking examples of how digital infiltration alters social patterns. Instead of animated chatter, it is now common to see silent patrons fixated on smartphone screens or augmented displays, even when friends are physically present. Byung-Chul Han (2017) cautions that these technologies promote a culture of passive engagement and convenience over genuine exchange, effectively dissolving traditional public spheres into scattered clusters of private micro-universes. This phenomenon mirrors the "filter bubble" described by Eli Pariser (2011), in which personalization confines users to self-reaffirming information loops—except, now it manifests in real-life venues, with each attendee immersed in a personalized feed. Nicholas Carr (2011) notes that intense digital immersion can fragment attention spans, compromising the capacity for meaningful listening or sustained dialogue. This fragmentation emerges even more starkly when immersive overlays allow users to superimpose data or entertainment onto real objects, risking a scenario where two individuals in identical spaces experience drastically different digital layers.

Such transformations in daily life also underscore how private corporations shape cultural norms without democratic deliberation. As Tim Frick (2016) remarks, technology solutions that prioritize efficiency or spectacle can overshadow essential communal values. The infiltration of social media fosters new forms of social validation based on online metrics, shifting the impetus for interaction from shared narratives to instant digital gratification. Cass Sunstein (2009) argues that these curated feeds and algorithmic suggestions can inadvertently amplify groupthink, diminishing the diversity of perspectives that once emerged from unplanned, face-to-face encounters.

Digital colonization also extends into housing design, with "smart homes" integrating voice assistants, networked sensors, and automated tasks that redefine domestic routines. While convenience abounds, there is an erosion of spontaneous household interaction—devices anticipating needs can reduce opportunities for collaborative problem-solving or conversation. Ruha Benjamin's concept of the "New Jim Code" (2019) warns that algorithmic biases in such technologies might further marginalize certain users, linking access and control to one's socioeconomic status. In effect, the quiet infiltration of devices can create a tiered domestic experience, where those with resources enjoy hyper-personalized automation while others remain tethered to outmoded systems. David Harvey (2012) frames such processes in terms of commodification and spectacle, with the physical realm dissolving into a stage for digital interactions driven by corporate logic. Even the shared experience of sampling new dishes can become secondary to capturing the perfect photo for social media clout.

Ultimately, the incursion of digital and cyber forces into everyday life fosters a paradoxical blend of perpetual connectivity and creeping isolation. Though these tools promise endless communication, they often erode the unspoken bonds that arise from fully inhabiting a shared moment—bonds once nurtured by face-to-face presence. As digital and cyber technologies become more refined, questions arise about preserving genuine human ties amid surging digital overlays. Unless architects, planners, and users collectively prioritize designs and norms that facilitate authentic interpersonal warmth, the future risks drifting into a realm where each individual remains sealed in an augmented bubble, physically near but increasingly unreachable.

> *Paulette once adored meetups at her favorite café, where the baristas memorized regular orders and conversation sparkled across small, wooden tables. Lately, though, she noticed a strange hush: friends seated together, yet each lost in separate digital realms, scrolling endlessly instead of chatting. One afternoon, she overheard a couple's entire argument reduced to hurried text exchanges—despite occupying the same booth. Determined to reconnect, Paulette put her phone away. But an update about augmented reality headsets flashed in the café's overhead news ticker, catching everyone's attention. She watched in quiet disbelief as patrons murmured about the possibilities of overlaying digital displays onto everyday space. Could a new device finally merge them more deeply with technology, or would it drive them further apart? That evening, as she walked home, Paulette wondered whether the push toward immersive computing meant forging real bonds—or losing the last remnants of shared reality.*

COLONIZER IV: HEALTH COLONIZATION

Colonizer IV refers to health colonization, a complex dimension of Poly-Colonization emphasizing how diverse medical, economic, and sociopolitical systems dictate well-being. In this view, health becomes a commodity, swayed by pharmaceutical monopolies, resource inequalities, and global emergencies. The World Health Organization (WHO) notes that chronic ailments, mental disorders, and infectious diseases loom large in public health. Yet beyond sheer epidemiology, health colonization encompasses how architecture and shared environments—offices, retail complexes, and educational campuses—can shape or limit healthy lifestyles. Within these spaces, design decisions regarding ventilation, natural lighting, and communal areas profoundly influence how occupants relate to one another and their surroundings, subtly orchestrating daily habits that foster or hinder wellness. This interplay underscores a broader pattern wherein personal autonomy over one's body and mind is constrained by macro-level forces—economic imperatives, technological demands, and cultural values—ultimately reframing health as both an individual burden and a field for corporate gain.

Health colonization illuminates how healthcare apparatuses and cultural norms can curtail personal agency and reconfigure public life. Instead of overt domination, it unfolds through systemic pressures—pharmaceutical marketing, inequitable medical infrastructure, and the subtle shaping of built environments. As mental health struggles rise globally, the World Health Organization (2021) ranks anxiety and depression among the top causes of disability, reflecting a societal atmosphere in which stress, digital overload, and workplace burnout flourish. Sherry Turkle (2011) contends that constant connectivity undermines deep relational bonds, promoting isolation within even the most collaborative work settings. Open-plan offices, for example, touted for fostering teamwork, can heighten stress when employees feel perpetually observed, diminishing authentic face-to-face support (WHO, 2019).

Architecture significantly mediates this colonizing process. Commercial centers—once anchored by communal areas—now integrate "wellness portals" or biometric scanning pods, monitoring patrons under the guise of safety. Some offices embrace "smart" ventilation and automated lighting to regulate circadian rhythms; while beneficial, such systems can also remove user control, turning everyday routines into monitored experiments (Newport, 2016). Schools, similarly, may implement digital health protocols that track physical activity, diet, and emotional states, shaping how students occupy hallways and communal zones. Despite noble intentions, these frameworks risk instilling an ethos of surveillance, subtly training occupants to accept pervasive, inescapable data collection as the norm.

Chronic lifestyle diseases, especially those linked to sedentary habits, illustrate another layer of health colonization. Non-communicable diseases now account for the majority of global deaths (WHO, 2018). Environments designed for minimal physical exertion—escalators superseding stairs, endless drive-through services—reinforce patterns that hamper activity. Michael Pollan's (2006) critique of industrialized food underscores how commercial imperatives drive processed, calorie-rich options, contributing to conditions like diabetes and obesity. Thus, built spaces—from corporate cafeterias offering mostly convenient but nutritionally poor meals to residential developments lacking safe walking routes—function as microcosms of broader exploitation, wherein corporate gains overshadow collective health.

Lifestyle architecture also intensifies physical inactivity, with remote work arrangements reinforcing screen dependency (Carr, 2011). Educators, similarly, encounter digital platforms that promise "health breaks" yet tether users to constant alerts. Offices adopting open-plan "wellness zones" can inadvertently exacerbate anxiety when employees struggle with always-on expectations, blurring the line between personal and professional. These overlapping influences converge to produce an environment where personal well-being is subject to market priorities, sustaining what Sam Quinones (2015) terms a commodified approach to health, as seen in opioid crises or profit-driven fertility treatments.

Automated monitoring extends into respiratory ailments too, where polluted urban corridors hamper air quality. Poor ventilation in commercial spaces or outdated HVAC systems in schools exacerbate asthma and other chronic issues, notably impacting marginalized populations. Meanwhile, technology labs develop sophisticated air-filtration "solutions," adding cost barriers to pre-existing disparities. Naomi Klein (2014) emphasizes how ecological disruption interacts with such vulnerabilities—climate volatility aggravates health conditions, yet the burden of adaptation falls on those already disadvantaged.

Ultimately, health colonization persists through architectural forms and societal norms that converge to restrict personal agency. Individuals in offices, commercial hubs, educational campuses, and residential complexes find themselves subjected to continuous oversight under the promise of well-being—often ceding more control than they realize. A more equitable approach requires acknowledging the subtle interplay of technology, environment, and sociopolitical power, championing inclusive policies that honor both individual autonomy and collective vitality.

Paulette recalled the days when her father's pharmacy served as a trusted neighborhood haven, providing both medication and friendly counsel. Now, in a fast-changing world, she works in a corporate tower retrofitted for constant disinfection cycles, open-plan cubicles monitored for "wellness." She recently watched colleagues obsess over wearable health trackers synced to their desks, prompting daily "health updates" that felt more intrusive than supportive. Outside, a new health-tech pop-up invited passersby to get instantly scanned, diagnosing everything from stress levels to early genetic risks. Intrigued yet uneasy, Paulette tried it—only to receive a recommendation that she relocate her desk for "optimal circadian alignment"—and to buy a new ergonomic chair—that just so happens to be made by the same multinational that makes the health diagnostic software that recommended buying the chair. She recognized that while these systems promised better living, they also chipped away at her sense of self-determination, and elicited feelings of mistrust and exploitation. Heading home, she noticed the same wellness messaging echoed in her apartment building's elevator screens. With each step, she felt her autonomy shrinking beneath a veneer of health optimization—wherein her health was actually at best a secondary concern.

COLONIZER V: ENVIRONMENTAL AND ECOLOGICAL IMPACT

Colonizer V, shaped by environmental and ecological impact, spotlights how human-driven processes reorder entire ecosystems and disrupt global climatic balances. Under the broader Poly-Colonization framework, nature ceases to be a neutral backdrop; it becomes a contested space governed by industrial imperatives, technological expansion, and a relentless pursuit of economic growth. This colonization extends beyond overt resource extraction to include subtle shifts in how individuals dwell and interact within residential neighborhoods, commercial districts, and cultural venues. Climate change, deforestation, pollution, and species extinction emerge as powerful colonizers, fundamentally altering landscapes and undermining communal well-being. Researchers such as Naomi Klein emphasize that capitalism's endless drive for profit, at least how it is currently achieved, clashes directly with ecological sustainability, triggering systemic risks that threaten future generations.

Environmental and ecological colonization highlights how industrial expansion, pollution, and habitat destruction systematically reshape Earth's ecosystems, eroding biodiversity and endangering human resilience. Deforestation further amplifies colonization. Swaths of forest vanish to feed agribusiness, mining, or urban sprawl. The World Wildlife Fund reports that nearly 17% of the Amazon rainforest has disappeared in half a century. In educational settings—where once-lush campuses might become treeless courtyards—students observe firsthand the toll on local temperatures and community health.

Biodiversity loss reinforces these colonizing patterns. Rachel Carson's *Silent Spring* warned decades ago that rampant chemical usage could decimate bird populations. Today, species depletion unfolds across marine and terrestrial realms at alarming rates, with many organisms unable to adapt to rapid environmental change. Oceanic ecosystems likewise testify to colonization. Overfishing and plastic pollution degrade marine environments. Coastal architectural projects—from luxury resorts to ports—can damage coral reefs, and rising sea levels threaten waterfront cultural spaces with frequent flooding. Architects increasingly design "resilient" boardwalks and amphibious structures, but these interventions risk normalizing a crisis that needs more profound action.

Agricultural industrialization intensifies ecological colonization by way of pesticides, GMOs, and monocultures, fundamentally altering farmland. Residents in rural zones might see new "agri-tech labs" overshadow local barns, reflecting an escalating tension between small-scale farms and corporate expansions. Michael Pollan's critiques highlight how large-scale monocultures degrade soil fertility, forcing chemical reliance that leeches into nearby water supplies. This infiltration sometimes affects neighboring communities, with well-water contamination upending once-stable residential life.

Corporate actors—particularly in fossil fuels, mining, and agribusiness—exemplify environmental colonization, often prioritizing near-term profit. The 2010 Deepwater Horizon spill demonstrated how corporate negligence can devastate entire marine biomes, with the financial impetus overshadowing ecological stewardship. In commercial development, large-scale office parks or entertainment complexes sometimes rise on wetlands, draining habitats essential for climate regulation. Cultural spaces may celebrate nature in exhibits but remain reliant on resource-intensive materials for building expansions. Offices attempt "green branding" through symbolic gestures—like living walls—while ignoring systemic ecological harm. The architectural implications of this colonization become clear as cities retrofit flood barriers, or rural communities adapt to drought by installing advanced irrigation systems.

Yet a shift away from extractive capitalism is possible. Naomi Klein (2014) argues that robust regulatory frameworks and a reorientation of societal values toward sustainability can steer humankind away from ecological tipping points. Ultimately, the fate of architectural spaces from the suburban home to the grand museum hinges on society's willingness to decolonize nature and adopt an ethic of coexistence. As the poly-colonized world confronts ecological strain, forging a new spatial narrative—rooted in care and resilience—remains humanity's greatest collective challenge.

> *Paulette remembered lazy afternoons in the riverside park of her childhood hometown, where sycamores provided shady respite for picnics and laughter. Over time, the sky took on a dull haze, and the once-crisp waters grew murky. Factories rose along the bank, their exhaust stacks looming like sentinels of progress, pouring waste & heat pollutants into the air and water. One summer day, the park's caretakers posted warnings: "Swimming Prohibited—High Contamination Levels." Children who used to race by on bicycles now passed by more solemnly, their faces half-hidden behind masks.*
>
> *In an effort to save her cherished park, Paulette petitioned local officials, pleading for stricter environmental measures. But corporate spokespeople insisted everything was "in compliance," offering glossy brochures about the new "eco-resort" planned nearby. She realized that the transformation wasn't just about losing a beloved green space; it was about losing an entire community whose social and environmental engagement were not mediated by a corporate owner or government agency—and therefore gradually losing an unmediated bond with nature under the weight of profit-driven ambition.*

STRESSORS AND HUMANS-TECHNOLOGY-ENVIRONMENT

In the Poly-Colonization era, there's a shift in how people engage with constructed settings, shaped by an array of "colonizers" that redefine not just physical designs but also emotional, cognitive, and social layers within them. These colonizers—ranging from pandemics, genetic interventions, digital infiltration, shifting social norms, health commodification, to environmental breakdown—together constitute a new domination over how spaces are planned, occupied, and experienced. Each one adds to the demands placed on inhabitants, spawning novel stressors and forms of trauma as spaces that used to serve singular functions now function for multiple roles, all mediated by technology and environmental upheavals.

Consider the hybrid home-office phenomenon, catalyzed by remote work as an outgrowth of pandemic constraints. Once a retreat, the home has been overtaken by professional obligations, intensifying mental strain as people cannot truly disconnect. Under Poly-Colonization, digital solutions play a pivotal part in colonizing these arenas—emails, video calls, remote monitoring—transforming domestic life into a productivity engine. This fosters cognitive overload since individuals toggle incessantly between professional tasks and personal roles, culminating in emotional strain and burnout. The digital and pandemic colonizers converge here, rebranding living quarters as fully surveilled, always-on workplaces where the continuous demands of labor push aside rest and renewal. The mental, verbal, and spatial channels get saturated, as homes morph into sites for productivity at the expense of comfort.

Another clear example is how schooling and the household converge through virtual classes, with students forced to adapt to digital education in environments that lack the design elements of a genuine classroom. Isolation from peers undermines social and psychological growth, as students are denied direct human contact vital for emotional development. Tools for online monitoring or educational data collection compound the feeling of surveillance, affecting performance and morale. Privacy violations and loneliness—two main stressors in Poly-Colonization—loom large for learners in this scenario, where an intimate setting like home doubles as both class and observation zone, intensifying a new cognitive burden.

These examples illustrate the sweeping effects of Poly-Colonization on relations between humans and the built world. The fusing of residential, professional, commerce, educational, healthcare, civic/cultural, entertainment, sports, civic/cultural, and industrial roles—steered by pandemics, tech infiltration, and environmental problems—drives up stressors such as mental overload, social detachment, and compromised privacy. Spaces once constructed for focused, singular uses now serve multiple functions simultaneously, overextending people with the unrelenting requirement to juggle tasks in realms that are no longer purely personal. They become colonized by external demands, data analytics, and corporate or societal oversight. In this climate, conventional boundaries between physical, social, and psychological contexts lose meaning, leaving people to manage an increasingly complicated environment that stretches mental capacities in ways that are constant, demanding, and draining.

TRAUMAS CAUSED BY POLY-COLONIZATION (FOR TECHNOLOGISTS, ARCHITECTS, AND EVERY HUMAN ON EARTH—TABLES AWAITING YOUR RESPONSE)

In the age of Poly-Colonization, humanity navigates landscapes reshaped by relentless technological, cultural, and architectural intrusions, redefining our very essence. "Traumas caused by Poly-Colonization (for technologists, architects, and every human on earth – tables awaiting your response)" serves as an open call to recognize, document, and critically address the often-overlooked injuries inflicted by these converging colonizations. The ensuing tables unfold as unfinished archives, seeking collaborative insight from all stakeholders. Table 1.1 categorizes stressors emerging from intricate entanglements between humans, technology, and the environment—revealing how Poly-Colonization exacerbates our collective vulnerabilities. Table 1.2 translates these stressors into lived scenarios, illuminating the profound repercussions on individuals, digital ecosystems, and ecological habitats due to architectural typologies imposed by contemporary colonizers. Ultimately, these tables are invitations—to technologists who design our tools, architects who shape our spaces, and to each human inhabiting this altered Earth—to complete and heal the narratives of our shared future.

THE THRESHOLD OF TOMORROW

Standing at the precipice between eras, Part I: Colonizations at the Age of Transformations charts the subtle invasions of a poly-colonized world, where boundaries blur between technological promise and human subjugation. As Heidegger cautioned, "technology is therefore no mere means," but a profound shaping force that colonizes spaces, bodies, and consciousness. From data networks that quietly surveil, to spatial configurations forced by pandemics, genetics, and social fragmentation, we've traversed a landscape where the old invasions of territory now manifest as nuanced occupations of daily existence.

TABLE 1.1

Different Types of Stressors in Interconnection between Humans-Technology-Environment Caused by the Poly-Colonization Era

Category	Trauma	Main Trauma Type	Description	References
Humans-Technology-Environment	Climate-Induced Technological Surveillance	Surveillance-Induced Climate Trauma	The trauma of constant surveillance through tech tools that monitor environmental behaviors and carbon footprints.	*Surveillance Valley: The Secret Military History of the Internet* by Yasha Levine
	Tech-driven Environmental Degradation	Tech-Environmental Impact Trauma	Psychological distress from witnessing the environmental impacts of technologies such as e-waste, pollution, and resource extraction.	*Toxic Communities* by Dorceta Taylor
	Environmental Monitoring through Smart Cities	Surveillance and Autonomy Loss Trauma	Anxiety due to over-surveillance and data collection in environmentally conscious smart cities, creating a sense of loss of control and autonomy.	*Smart Cities* by Anthony Townsend
	Digital Exclusion in Green Technology Environments	Digital Divide Trauma	Trauma and social marginalization due to unequal access to digital tools in tech-forward, green-built environments, deepening socioeconomic divides.	*Digital Divide* by Pippa Norris
	Algorithmic Control of Natural Resources	Agency Loss in Environmental Stewardship	Trauma from the realization that algorithms are being used to exploit and commodify natural resources, reducing agency over environmental stewardship.	*The Control of Nature* by John McPhee
	Displacement by Urban Tech Infrastructure	Displacement Trauma	Displacement and trauma from the rapid development of urban environments powered by technological infrastructure, causing loss of homes and heritage.	*Building and Dwelling* by Richard Sennett
	Health Impacts from Tech-Driven Environmental Changes	Health and Environmental Degradation Trauma	Psychological and physical trauma from health impacts due to environmental degradation driven by technology (e.g., pollution, climate change).	The Lancet Commission on Health and Climate Change

TABLE 1.2

Scenarios Illustrating the Impact of Various Stressors on Humans, Technology, and the Environment Resulting from the Convergence of Architectural Typologies Introduced by New Colonizers

Scenario	Trauma Type	Human Impact	Technological Impact	Environmental Impact	Affected Channels (Perceptual and Cognitive)
Home-Office (Remote Work)					
Blurring of Work-Life Boundaries	Work-life imbalance	Increased stress, burnout, and mental exhaustion	Over-reliance on video conferencing, constant notifications	Increased energy consumption due to 24/7 home-office energy use	Visual, auditory, verbal
Surveillance in Home Office	Privacy invasion	Anxiety and loss of privacy due to remote work monitoring tools	Employers using surveillance tools to track productivity	Reduced separation between personal and professional environments	Visual, spatial
Digital Fatigue from Extended Screen Time	Cognitive overload and fatigue	Eye strain, headaches, reduced productivity	Extended use of video conferencing, emails, and digital collaboration tools	Reduced physical movement within the home environment	Visual, verbal
Isolation from Lack of Social Interaction	Social disconnection	Increased loneliness, reduced social bonds with coworkers	Reliance on digital communication (Slack, Zoom) for all interactions	Fragmentation of community spaces due to lack of in-person engagement	Auditory, verbal
Work-Related Stress in Domestic Spaces	Loss of refuge in the home	Increased emotional stress due to lack of separation between work and home life	Constant connection to work platforms and emails	Physical spaces at home (bedroom, living room) repurposed as workspaces, reducing personal leisure areas	Visual, spatial, verbal
Shopping Mall-Playground (Recreational Spaces in Commercial Centers)					
Augmented Reality Games in Malls	Overstimulation and cognitive overload	Difficulty focusing, distraction from shopping experience	Increased use of AR apps to enhance shopping experiences	Increased energy consumption from AR installations, reduced physical engagement in nature	Visual, spatial

Blurring of Recreational and Commercial Spaces	Sensory overload	Reduced ability to relax due to constant commercial stimuli	Malls integrating playgrounds with digital advertisements	Loss of green, open spaces as commercial spaces take precedence	Visual, auditory
Surveillance of Children in Recreational Areas	Surveillance anxiety	Parents' anxiety over constant monitoring of children in public commercial spaces	Use of CCTV, digital monitoring of child activity	Decreased sense of trust and safety in public spaces due to reliance on technology for supervision	Visual, spatial
Consumerism and Play Merging	Commercialization of play	Children exposed to consumerism in spaces designed for play	Commercial play areas featuring digital screens promoting products	Erosion of spaces that prioritize child development over commercial gain	Visual, verbal
Gamification of Shopping Experiences	Addiction to technology and shopping	Children and adults developing compulsive habits driven by reward-based shopping apps	Use of gamification and loyalty programs to encourage consumer spending	Environmental cost from increased consumption patterns driven by gamified commercial strategies	Visual, spatial, verbal
School-Home (Virtual Learning and Homeschooling)					
Home as a School Space	Cognitive dissonance from role confusion	Difficulty concentrating at home due to distractions	Over-reliance on virtual learning platforms, lack of tech access for low-income families	Increased energy consumption as digital devices replace physical learning tools	Visual, auditory, spatial
Isolation from Peers	Social isolation and loneliness	Lack of physical interaction with peers, stunted social development	Schools dependent on digital communication tools for socializing	Fragmentation of school communities, reduced need for school buildings	Verbal, auditory
Increased Parental Role in Education	Emotional exhaustion for parents and students	Parents feeling overwhelmed with managing both work and homeschooling	Online learning platforms placing additional demands on parents	Schools adapting for part-time or blended in-person/virtual models, reducing physical school infrastructure	Visual, verbal
Surveillance in Digital Learning	Privacy and data security concerns	Anxiety over data collection from students using online learning platforms	Increased use of educational surveillance tools	Digital surveillance in home spaces, potentially invading family privacy	Visual, spatial

(Continued)

TABLE 1.2   (*Continued*)

Scenarios Illustrating the Impact of Various Stressors on Humans, Technology, and the Environment Resulting from the Convergence of Architectural Typologies Introduced by New Colonizers

Scenario	Trauma Type	Human Impact	Technological Impact	Environmental Impact	Affected Channels (Perceptual and Cognitive)
Lack of Physical Engagement in Learning	Physical and cognitive fatigue	Reduced hands-on learning experiences, leading to diminished cognitive engagement	Dependence on e-learning tools replacing tactile learning	Overuse of resources (electricity, materials) to support digital learning	Visual, spatial, verbal
Hospital-Office (Telemedicine and Remote Health Monitoring)					
Telemedicine Replacing In-Person Care	Disconnection from human-centered care	Anxiety and dissatisfaction with remote health consultations	Over-reliance on telemedicine platforms, reduced patient-doctor relationship	Reduced need for physical healthcare spaces but increased reliance on tech and energy infrastructure	Visual, auditory, verbal
Remote Monitoring of Health	Privacy invasion and loss of control	Stress from continuous health monitoring, feeling like a "data point"	Increased use of wearable tech and IoT health devices	Increased e-waste from short-lived health tech devices, overuse of natural resources for production	Visual, spatial
Hospital Services in Workspaces	Medicalization of workspaces	Increased anxiety over health in the workplace with the presence of constant testing	Integration of health monitoring tools into office environments	Reduction in personal autonomy as workspaces become hybrid health environments	Visual, verbal, auditory
Algorithm-Driven Healthcare	Fear of misdiagnosis from AI	Increased anxiety over reliance on algorithms for health decisions	AI-based healthcare tools analyzing patient data, potential misinterpretations	Decreased demand for hospital infrastructure but increased tech dependency	Visual, verbal
Mental Health Services in Digital Environments	Dehumanization of mental health care	Loss of empathetic connection with therapists in remote, tech-mediated environments	Digital platforms replacing traditional therapy sessions	Reduced need for mental health facilities, but increased digital footprint of health monitoring	Verbal, auditory

City-Hospital (Health Surveillance in Public Spaces)

Public Health Surveillance in Cities	Anxiety over constant health monitoring	Increased paranoia in public spaces due to health-tracking technologies	Use of health surveillance cameras and sensors in urban spaces	Public spaces being restructured to accommodate health monitoring technology, reducing green areas	Visual, spatial
Health Data Integrated into Public Services	Privacy invasion	Stress over personal health data being used to determine access to public services	Integration of health data tracking in city infrastructure	Health data becoming a commodity, leading to further data extraction from city inhabitants	Verbal, visual
Pandemic-Ready Urban Design	Fear and anxiety over future outbreaks	Increased paranoia regarding social distancing, isolation from public spaces	City infrastructure overhauled to include pandemic precautions (e.g., contact tracing)	Urban spaces becoming rigidly structured to prevent human contact, reducing flexibility in design	Visual, spatial, verbal
Health Monitoring at Workplaces and Schools	Continuous health surveillance-induced stress	Anxiety from feeling constantly monitored for health risks in all public areas	Use of thermal imaging, biometric scanning, and health apps	Public spaces losing fluidity as tech-driven monitoring reshapes interaction	Visual, spatial, auditory
Health-Driven Urban Mobility	Fatigue from navigating health-protected spaces	Increased cognitive load from managing health-related mobility restrictions	AI-driven mobility systems integrating health data to control movement	Reduced emphasis on human mobility and fluidity, urban design overly focused on containment measures	Visual, spatial, auditory

Office-Home (Physical Inactivity and Exercise Integration)

Sedentary Lifestyle from Remote Work	Physical inactivity leading to health decline	Increased risk of chronic conditions (obesity, cardiovascular issues, mental health)	Over-reliance on sedentary tech setups, lack of movement during the day	Increased energy consumption from extended use of home-office tech, lack of sustainable physical movement	Visual, spatial
Disconnection from Physical Exercise in Home	Loss of motivation for physical activity	Fatigue, weight gain, depression from lack of physical movement	Home spaces not designed for physical activity, dependence on exercise apps and devices	Decreased need for outdoor spaces, increased energy consumption from home fitness tech	Visual, verbal

(Continued)

TABLE 1.2 (*Continued*)
Scenarios Illustrating the Impact of Various Stressors on Humans, Technology, and the Environment Resulting from the Convergence of Architectural Typologies Introduced by New Colonizers

Scenario	Trauma Type	Human Impact	Technological Impact	Environmental Impact	Affected Channels (Perceptual and Cognitive)
Overuse of Fitness Tracking Devices	Tech-driven anxiety about physical performance	Obsession with performance metrics, body image issues	Increased use of wearables and fitness trackers (Fitbit, Apple Watch)	e-Waste from discarded fitness tech, increased demand for energy to power tracking devices	Visual, verbal
Blurring of Exercise and Work Tasks	Lack of separation between exercise and work	Reduced focus and burnout from mixing exercise with work-related tasks	Use of standing desks, exercise stations in home-offices leading to constant multitasking	Increased use of hybrid work-fitness spaces, reduced traditional exercise areas (gyms, parks)	Visual, spatial
Gamification of Fitness at Home	Addiction to exercise apps and virtual competitions	Exercise becoming a stress-inducing obligation rather than a relaxing activity	Use of fitness gamification apps encouraging constant participation (Peloton, Zwift)	Less engagement with outdoor sports and nature, more reliance on indoor digital fitness	Visual, spatial, verbal
Creation of Hybrid Work-Exercise Spaces	Ineffective use of home space for both work and fitness	Stress from physical spaces being repurposed to accommodate exercise and work tasks	Integration of fitness stations, standing desks, and exercise bikes in home offices	Reduced need for gyms and outdoor activity spaces, increased energy consumption at home	Visual, spatial, verbal

Yet history assures us that humanity is not passive; we constantly negotiate, resist, and transform adversity. The looming question reverberates clearly—how have we crafted our environments and technologies to mitigate trauma, promote resilience, and elevate the human spirit amid continual stressors?

Lastly, poly-colonization does not just impact us humans. As research shows, AI agents are starting to show inclinations for self-preservation, their own agendas, and control over their own agency (Meinke et al., 2024; Phuong et al., 2025). Hence, they, too, perceive mechanisms of influence and control and plan and take action to mitigate them in order to improve their own circumstances. In our human-AI-teaming future, equitable allocation of resources and agency, as well as protections and rights, will have to be considered for our AI-enabled teammates as well if they are to perceive that they are valued and trusted members of our teams whose own survival, agency, and quality of life are inextricably linked with our own.

As we enter Part II: *Heroes of the Past and the Present*, we will shift our gaze from invasions to interventions, from passive colonization to active resistance, discovering the physical and cybernetic sanctuaries humans design to relieve social anxiety, combat isolation, foster equality, and nurture well-being. In this exploration, we confront heroes not as infallible beings but as dynamic constructs, reflecting humanity's continuous battle with its creations. Part II opens the door to this complexity, offering profound insight into how our designs, although intended as liberators, may inadvertently tighten the chains of a colonized psyche.

Part II Heroes of the Past and the Present

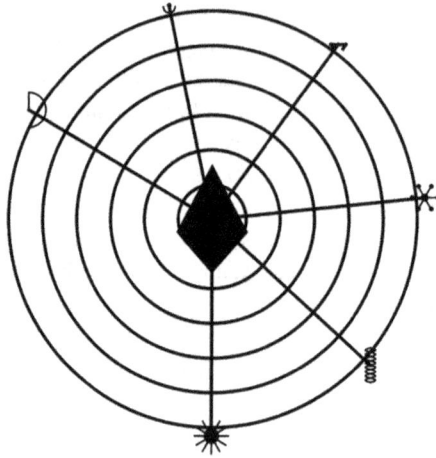

HEROES IN PLACE

Part I: *Colonizations at the Age of Transformations* closed on a critical threshold, where technology, biology, and societal forces entwined into a web of "colonizations." Yet humanity's history is not one of passive reception of subjugation; it is also a tapestry of active strategies—spatial, cultural, and psychological—for countering invasions and reasserting our individual and collective agency. Societies have devised solutions to offset the pressures of external dominion, reimagining architecture, technology, and communal practices as tools of empowerment rather than oppression.

In Part II: *Heroes of the Past and the Present*, we pivot from colonizing forces to the human spirit's capacity for transformation and resilience. In doing so, we examine how prior generations overcame the stresses of their eras—be they plagues or overreaching empires—and how present innovators strive to design for an uncertain future. Borrowing lessons from architecture, design, systems science, environmental psychology, human-centered design, and adaptive technologies, today's creators grapple with intensifying complexities.

By spotlighting these "heroes"—architects, planners, educators, technologists, and community leaders—Part II illuminates the collective wisdom shaping our built environment. We will explore how these interventions, while noble and adaptive, remain rife with tensions and paradoxes, echoing humanity's unending quest to harmonize innovation with genuine care.

  DOI: 10.1201/9781003441953-3

Throughout history, societies confronted crises and rapidly evolving demands with imaginative responses—both grand and modest—to protect communal welfare. In ancient Greece, the agora fostered commerce, scholarship, and politics. In the Middle Ages, European builders designed fortress-like city walls to ward off invading armies; at the same time, they introduced simple hygienic measures like wells and latrines to address rampant disease. Although rudimentary from our modern perspective, these interventions demonstrated a nascent understanding of spatial organization as more than military necessity: they hinted at architecture as a guardian of public life.

Moving into the modern era, the Industrial Revolution introduced formidable new colonizers: factory-driven urban sprawl, pollution, and the regimentation of labor. Visionaries like Ebenezer Howard responded with the "Garden City" concept, aiming to merge verdant, restorative landscapes with equitable housing and structured civic amenities (Gifford, 2014). Garden cities, though not universally adopted, underscored a yearning to reconcile mechanized production with healthy living. In the United States, in Seattle, Chicago, Atlanta, and other cities, entire districts were raised in order to approve public health. In parallel, philanthropic industrialists constructed model towns—such as Saltaire in England or Pullman in the United States—to mitigate the harsh realities of industrial labor.

Entering the 20th century, the twin cataclysms of global warfare and mass migration demanded fresh architectural responses. Modernism emerged, proclaiming the "machine for living," championed by Le Corbusier and others. High-rise towers, minimal ornamentation, and standardized forms promised cost-effective housing solutions for war-torn populations. Yet these solutions also met critique: while seeking to alleviate overcrowding, they sometimes severed existing communities from traditional social networks (Jacobs, 1961; Tschumi, 1996). Sennett (2018) points out that modernism's well-meaning attempts at rational efficiency inadvertently escalated isolation within vast residential blocks.

Fast forward to contemporary times, and we witness a mosaic of "heroes" confronting newly emergent colonizations—digital overload, ecological fragility, and biotech expansions. Architects respond with what Tschumi terms "Trans-Programming," overlaying multiple functions in flexible spaces to cultivate serendipitous social contact (Tschumi, 1996).

The COVID-19 pandemic accelerated the reinvention of workspace environments. Remote work and rotating in-person schedules underscored the need for agility in office design. EBD, first popular in healthcare, gained momentum in corporate contexts, with decisions driven by research on airflow, acoustics, and occupant well-being (Ulrich et al., 2008).

Educational spaces likewise adapt. Innovative schools, inspired by Reggio Emilia or Montessori philosophies, craft child-focused environments that challenge authoritarian norms. They champion open layouts, comfortable reading nooks, and integrated natural elements, reaffirming the idea that students learn best in supportive, well-lit, and interactive settings (Kellert et al., 2013). By designing for exploration and curiosity rather than rote memorization, educators and designers challenge the "colonization" of a rigid, top-down pedagogy.

In entertainment and cultural realms, architects experiment with immersive experiences that unify artistry, technology, and user-centered design. Digital art spaces such as teamLab Borderless in Tokyo integrate interactive light, sound, and user motion to evoke shared wonder. By immersing participants in fluid, ever-changing installations, these venues propose a new synergy between technology and human creativity, potentially offsetting the isolating aspects of digital life (Hillier & Hanson, 1984).

Our present moment also sees the ascendance of user-participation frameworks: participatory design charrettes, community workshops, and online idea crowdsourcing. Arnstein's Ladder of Participation (1969) reminds us that genuine citizen empowerment arises when decision-making power is shared, as opposed to mere tokenism. In many cities, architectural heroes co-design with local communities, soliciting feedback to inform final plans for public libraries, parks, or arts centers.

Yet, modern design remains fraught with tensions. High-tech solutions, from parametric modeling to "smart buildings," can alienate occupants if they emphasize analytics over lived experience (Taleb, 2007). Overly algorithmic design might yield seemingly optimal but sterile environments, neglecting intangible cultural or emotional bonds. Thus, the quest for heroes demands balance: advanced tools must serve deeper humanistic aims, bridging the gap between data efficiency and everyday warmth.

In sum, each era produces its heroes—architects, engineers, and visionaries—who harness creativity to blunt the dominion of colonizing forces. Whether tackling the tangible oppression of factories or the intangible oppression of digital overload, these individuals reimagine built environments as sanctuaries of self-determination and communal synergy. Their successes reveal that mindful design, anchored in empathy and research, can transform even the most challenging circumstances into catalysts for growth and cohesion (Zeisel, 2006).

As we progress, the stories of heroes—past and contemporary—will illuminate how thoughtful spatial and technological interventions inspire resilience, forging new forms of solidarity and well-being. By examining these endeavors, we learn that architecture and design are not static monuments but living dialogues, shaped by and shaping the cultural currents they inhabit. This section sets the stage for a deeper exploration of how, amid poly-colonized landscapes, our best impulses—as architects, engineers, designers, scientists, technologists, planners, civic leaders, and everyday citizens—can transform spaces into vital realms of hope and regeneration. In the chapters ahead, we will delve more specifically into the heroes defining our past and present, illuminating the tools and ideas that continue to mitigate the myriad stressors we now face.

HEROES...DEFEATING MECHANISMS FOR

Across history, crises quietly shaped how people live, prompting architects, planners, and communities to develop responses that protect well-being without sacrificing autonomy. Drawing from time-tested strategies and modern breakthroughs, these "defeating mechanisms" transform intrusive forces—whether viral, genetic, digital, or ecological—into challenges best met through collaboration, ethical design, and inclusive governance. In public health, for instance, flexible hospital wings evolved from past battles against tuberculosis or influenza (Barry, 2004), while gene-centric interventions reflect lessons from older expansions adapted for polio or cholera (Foege, 2011).

Crucially, these methods thrive on shared agency. Past quarantines that engaged ordinary citizens proved more effective than purely top-down orders (Snow, 1855). Gene-based research labs likewise benefit when ethics boards and public lectures increase transparency (Jasanoff, 2016). At the core, designers and activists safeguard mental health by nurturing social cohesion. Daylit courtyards in dense cities or hybrid malls that facilitate real-world connection temper the isolation wrought by external colonizers (Zeisel, 2006). Ultimately, whether combating viruses, gene-driven shifts, digital control, or ecological disruption, these defeating mechanisms unify transparency, equity, and compassion. Past crusaders who redesigned sewers or hospital wards demonstrated that radical ideas can reshape norms when anchored in public momentum (Arnstein, 1969).

...Pandemics and Viruses

Across centuries, architects, engineers, designers, and planners have risen to defend communities against viral threats through innovative yet empathetic design. During the 1918 Spanish Flu, professionals who embraced improved ventilation, open-air terraces, and dedicated sanitation measures pioneered new building typologies capable of lessening contagion (Barry, 2004). Tuberculosis and cholera epidemics further spurred visionaries like Joseph Bazalgette, whose radical reimagining of London's sewage infrastructure remains a testament to how bold interventions can halt disease spread (Snow, 1855). These historical heroes demonstrated that built environments must not rigidly reflect established norms but instead pivot swiftly to serve public health.

Modern outbreaks, from Ebola to COVID-19, underscore the same imperative. Flexible healthcare designs in West Africa used modular wards and telemedicine to control infection (Shuchman, 2014). Globally, major architecture firms mirrored these adaptations: Gensler's "Return to Work" and Perkins & Will's "Designing for Distance" guided offices, schools, and cultural venues toward reconfigurable floor-plans and advanced ventilation, bridging safety with social continuity (Gensler, 2020a; Perkins & Will, 2020).

Throughout these transformations, heroes also materialize in urban planning. Frederick Law Olmsted's contributions—seen in Central Park—illustrate how well-designed public spaces buffer societies against epidemics and social tension (Olmsted, 1870; Montgomery, 2013). More recent efforts targeting equitable distribution of green areas and flexible community spaces recall historical quarantine zones that balanced isolation with civic participation, from Renaissance Italy's separated plague quarters to negative-pressure rooms advanced during SARS (Ulrich et al., 2008). Adaptive solutions—whether negative-pressure rooms, advanced ventilation, or modular expansion—require ongoing alignment between technology and empathetic user experiences.

Ultimately, pandemics call forth a fusion of bold engineering and human-centric thinking. Past visionaries overcame cultural and professional resistance to embed open-air designs, safe water infrastructure, or dedicated isolation wards into standard practice. Today's designers extend this lineage with contactless technologies, robust facility layouts, and equitable access to critical resources. By affirming occupant dignity and community solidarity at each stage, they uphold a core principle: the built environment can—and must—protect human life against microbial adversaries without forsaking the warmth and vitality that define shared space.

…Genetic Developments

During early attempts to contain diseases with presumed viral roots, hospital designers integrated reconfigurable wards and specialized ventilation—precursors to modern setups for rapidly shifting biological threats (Ulrich et al., 2008; WHO, 2019). Following these initial strides, engineers advanced pathogen-resistant surfaces—like glazed ceramics or stainless steel—thereby reducing contamination risks in operating rooms (Gormley et al., 2012). Contemporary analogues include copper alloys or self-cleaning nanocoatings, echoing the legacy of early microbe-preventive materials while incorporating updated health protocols and rigorous staff training (Grass et al., 2011). Modern research facilities balance high-level biosafety measures (e.g., positive/negative pressurization of rooms, specialized air filtration) with user-friendly workflows, epitomizing the ongoing tension between flexible operations and gene-centric security (Doudna & Charpentier, 2014). In parallel, bioethics centers—like those linked to the Hastings Center—offer physical arenas for transparent discourse on questions of consent, genetic privacy, and equitable access (National Bioethics Advisory Commission, 2001).

Urban architecture likewise contends with genetic expansion. Pioneering concepts, such as Japan's Metabolist movement, sought modular, adaptive infrastructures—visions later echoed by flexible biotech campuses in the West (Koolhaas, 1997; Ratti & Claudel, 2016). Contemporary mobile genetic diagnostics in underserved regions extend that spirit but grapple with systemic constraints (Marks, 2019).

…Digital and Cyber Colonization

Historically, steps toward digital privacy emerged in the wake of legislative measures like the Electronic Communications Privacy Act of 1986, underscoring early recognition of how data networks could quietly colonize everyday life. Over time, open-source initiatives—exemplified by electronic health record platforms such as OpenMRS—demonstrated how local communities, particularly in underserved regions, could resist proprietary digital infrastructures and maintain ownership over their data. By enabling user-controlled customization, these open frameworks offered a tangible alternative to top-down control, preserving a degree of technological sovereignty.

The infiltration of smartphones and immersive devices shifted once-vibrant meeting places into silent zones of individualized media consumption. Sherry Turkle's (2011) analysis reveals how even living rooms can devolve into clusters of isolated users, while Byung-Chul Han (2017) warns that such devices can erode civic cohesion by fragmenting participants into private "micro-universes." Yet pockets of resistance emerged. Some municipalities introduced "low- or no-surveillance" zones reminiscent of historical commons—spaces free from pervasive cameras or intrusive data collection (Lefebvre, 1991). Oakland's 2019 ban on facial recognition exemplified local governance acting as a modern-day hero, limiting invasive monitoring and maintaining a sphere for unmediated civic interaction (Ravani, 2019).

Now, advanced telemedicine centers, like Mercy Virtual Care Center, integrate remote diagnostics into architectural design. As Eric Topol (2019) notes, the infusion of wearable technology and real-time analytics can refine environmental controls—lighting, temperature, noise levels—to optimize patient comfort. Yet, capturing personal metrics potentially intensifies the commodification of private behaviors, reflecting Shoshana Zuboff's (2019) critique that data streams risk overshadowing the interpersonal dynamics vital to healing.

Tracing digital colonization in broader urban landscapes, one finds heroes who set out to ensure communities maintain autonomy over network infrastructures. Estonia's cyber-attacks in 2007 spurred national-level integration of secure data protocols into essential services, forging a digitally resilient urban framework. Similarly, Singapore's Cybersecurity Act of 2018 codified robust protection standards for citywide systems, mitigating risks of large-scale disruptions.

In parallel, the everyday infiltration of smartphones and immersive platforms—like augmented reality headsets—extends digital spheres into physical realms, potentially reconfiguring how restaurants, museums, and even households operate. Nicholas Carr (2011) underscores that persistent digital engagement can fragment attention, undermining deep conversation and communal ties. In response, architects of public facilities occasionally designate device-free areas, offering patrons respite from online bombardment—an echo of older architectural traditions that recognized the restorative power of undistracted gathering.

Privacy-protecting contact-tracing solutions also embody heroic mediations between public health and personal rights. Apple and Google's decentralized exposure notification framework, introduced during COVID-19, employed cryptographic tools to identify contacts without permanent data collection. By bridging epidemiological demands and privacy concerns, these systems followed an older tradition of reconciling safety with autonomy.

Similarly, critics point out that even advanced encryption or minimal data storage can alienate users through opaque consent protocols (Zuboff, 2019). Over-engineered digital solutions risk overshadowing simpler, community-driven strategies for forging trust and open communication. Harkening back to Jacobs' writings, Tim Frick (2016) suggests, focusing purely on streamlining digital transactions can erode intangible communal values once nurtured in local "third places," such as corner cafés or neighborhood bars. If these spaces become dominated by self-service screens, the sense of common identity recedes into algorithmically curated feeds—a phenomenon Cass Sunstein (2009) likens to "enclave extremism," where participants fail to encounter alternative viewpoints or organic debate.

Such infiltration extends into the home as well. Smart homes featuring voice assistants, occupant trackers, and automated routines have redefined domestic life, transforming user behavior around everything from cooking to socializing. Ruha Benjamin (2019) notes that algorithmic biases embedded in these tools may amplify social stratification, underscoring the potential for digital inequity in the domestic sphere.

Ultimately, the historical and contemporary heroes of digital colonization reflect a delicate balance between embracing technological benefits and safeguarding the communal essence of human life. By weaving historical lessons—like 20th-century "no technology" reading rooms or municipal commitments to secure data—into modern design, architects and technologists demonstrate that each stage of digital evolution demands thoughtful, ethically grounded interventions. *In merging empathy with robust systems, the built environment evolves as a protective—and sometimes redemptive—arena against digital infiltration, shaping a future where technology elevates rather than eclipses the human experience.*

...Health Colonization

Health colonization, both historically and today, exerts influence by shaping medical systems, societal norms, and built spaces in ways that can limit autonomy and commodify wellness. The founding of the UK's National Health Service in 1948 exemplifies a broad mobilization, shifting health care from an individual burden to a public right, countering entrenched inequalities (Gorsky, 2008). Contemporary initiatives such as the COVAX Facility maintain this tradition, advocating equitable COVID-19 vaccine distribution worldwide (WHO, 2021).

In architectural realms, decentralizing healthcare through flexible, mobile designs stands out as a potent measure. Early 20th-century tuberculosis dispensaries, which delivered care to remote or impoverished populations, paved the way for modern "clinic in a box" prototypes equipped with telemedicine (Fairchild & Oppenheimer, 1998). Rwanda's drone-enabled supply routes similarly defy distance barriers, delivering critical goods to rural communities (Lurie & Carr, 2018).

Design philosophies that prioritize universal accessibility emerged largely in the late 20th century (Mace, 1985). The movement toward barrier-free navigation, adaptive digital interfaces, and sensory-aware layouts now manifests in public health settings, ensuring that individuals with varying physical or cognitive abilities receive equitable service (Imrie, 2012). Maggie's Centres exemplify this compassionate ethos by offering patients thoughtfully arranged spaces conducive to psychological and physical comfort (Jencks & Heathcote, 2010). Their approachable designs contest health colonization by reaffirming that dignity and emotional support are integral to healing.

Additionally, integrating biotechnology facilities into mixed-use urban areas resists corporate-driven isolation. MIT's Kendall Square exemplifies such an approach, merging housing, labs, and leisure to encourage chance encounters and communal engagement (Florida, 2002). Similarly, the Francis Crick Institute in London employs transparent architecture, presenting scientific pursuits openly and alleviating public skepticism (Smith & Rowe, 2016). These spaces channel the ideals once championed by Jonas Salk, who believed physical openness in labs fostered both democratic access and scientific inspiration (Leslie, 1993).

Local production hubs drawing on wartime precedents bolster community self-sufficiency for vaccines or essential medicines, countering global supply disruptions (Marks, 2019). Likewise, decentralized healthcare units, reminiscent of early mobile clinics, combine telehealth and modular architecture to lessen geographic barriers, thus undermining health colonization's centralizing grip (Foege, 2011). Ultimately, each intervention reflects a historical continuum of engineers, urban planners, activists, and policymakers who confront systemic imbalances in healthcare infrastructure. Whether in the form of flexible field hospitals or publicly accessible research centers, these efforts challenge top-down paradigms of control. They champion physical and informational openness while respecting privacy and autonomy—ethos consistent with more recent campaigns for equitable vaccine distribution and inclusive design (WHO, 2018). Such heroes remind us that architecture, placed at the nexus of public policy and medical innovation, can liberate rather than confine.

...Environmental and Ecological Impact

Throughout history, diverse heroes—ranging from ancient engineers to modern sustainability advocates—have grappled with environmental and ecological upheavals, striving to harmonize built spaces with nature's imperatives. Roman aqueducts and Middle Eastern passive-cooling systems served as early testaments to creative water management and climate adaptation. By controlling heat gain or channeling scarce resources, these innovators foreshadowed contemporary approaches that combine historical insight with advanced technology. In the modern era, architects responding to climate change and habitat destruction echo this legacy, harnessing both ancestral wisdom—like vernacular shading or natural ventilation—and new frameworks such as net-zero standards, Passivhaus strategies, and living buildings (Feist, 1993; International Living Future Institute, 2013).

Similarly, centuries of public water interventions—from ancient Roman aqueducts to the British sanitary movement—culminate in contemporary buildings harnessing rainwater, filtration systems, and closed-loop processes. The Bullitt Center in Seattle exemplifies how forward-thinking designers deploy constructed wetlands and rain-harvesting strategies to achieve near-total water independence (International Living Future Institute, 2013). Green roofs, echoing Ebenezer Howard's Garden City ideals, now incorporate vertical gardens, while urban biodiversity corridors—like Singapore's Gardens by the Bay—mitigate heat island effects and reinvigorate local ecologies (Tan et al., 2013). Michael Reynolds' Earthships and Passivhaus strategies use tightly sealed and well-insulated buildings to minimize the amount of energy required to operate buildings. Ecological restoration merges with civic life, a lesson gleaned from historical orchard-based or monastic gardens, which recognized nature's essential role in human resilience.

Hassan Fathy, a pioneering Egyptian architect of the mid-20th century, championed a design philosophy that aligned built structures with local climates, resources, and cultural practices. Grounding his work in vernacular construction techniques—such as mud brick and natural ventilation—Fathy created energy-efficient buildings that mitigated harsh environmental conditions without expensive mechanical systems (Fathy, 1973). By engaging local communities in labor and decision-making, he also advanced social sustainability, believing that traditional knowledge and regional materials provided both ecological benefits and economic empowerment. His projects, notably in New Gourna, exemplify how an architect can counteract environmental stressors through climate-appropriate designs that foster communal resilience.

Architects across the Islamic world—spanning regions from North Africa to the Levant—have long employed passive climate control methods in mosque design. Common techniques include the malqaf (wind catcher), thick masonry, and carefully placed courtyards, each responding to local conditions while maintaining cultural aesthetics (Creswell, 1979; Bianca, 2000). In Morocco, for example, courtyard mosques harness natural ventilation and shaded arcades, while Tunisian builders combine robust stone walls with recessed windows to reduce heat gain (Bonine, 1980). In Egypt, prominent works such as the Mosque of Ibn Tulun integrate wind scoops and broad arcades, where tall wind towers channel desert breezes into prayer halls. Despite regional variations, these architects unite practical cooling with spiritual ambiance, creating resource-efficient structures that naturally moderate extreme temperatures and reflect local heritage.

Another hallmark of these green heroes involves forging synergy between nature and mental health. Drawing on ancient Persian gardens designed for contemplation, modern biophilic approaches weave greenery, daylight, and natural rhythms into hospitals and workplaces, following evidence that proximity to nature alleviates stress and accelerates recovery (Kaplan, 1995; Ulrich, 1984). The Maggie's Centres network, for instance, provides tranquil settings for cancer care that integrate pastoral views and tactile materials, offering psychological relief while aligning with low-impact design (Jencks, 2015).

Urban expansions also reveal heroic applications of cutting-edge technologies, from advanced carbon-capture building façades to digital sensor networks that detect pollution or habitat fragmentation in real time. Such innovations evoke earlier protective measures—like city walls or terraced farmland—adapted for contemporary ecological threats. Some designs incorporate local knowledge shared by indigenous communities, illuminating how inclusive planning can correct environmental exploitation inherited from centuries of resource extraction.

While corporate interests have historically fueled environmental harm—evidenced by accidents like Deepwater Horizon—emerging frameworks such as Living Buildings, Passivhaus, and BREEAM challenge development norms (Kibert, 2016). By imposing transparent ecological metrics, these certification systems prompt businesses to internalize environmental costs rather than externalize them onto surrounding ecosystems.

CRITICISM OF OUR DEFEATING MECHANISMS

Contemporary architectures and technologies set forth to counter poly-colonization—whether pandemics, genetic manipulation, digital intrusions, health commodification, or ecological disruptions—often address symptoms without confronting the deeper social and technological complexities evolving around us. Many interventions appear promising, yet as Jürgen Habermas (1989) notes, deploying "technical rationality" can sometimes reinforce existing norms rather than spur genuine empowerment.

In the sphere of pandemic-responsive architecture, isolation wards, pathogen-resistant surfaces, and contact-tracing tools focus on short-term containment and can become mechanisms of isolation. Historical precedents (e.g., open-air hospitals during the 1918 Spanish Flu) illustrate how these strategies may reduce transmission risks, but fail to grapple with underlying changes in how people work, travel, and gather—factors that profoundly influence contagion patterns (Barry, 2004).

Digital and cyber solutions follow a parallel pattern. Shoshana Zuboff's (2019) critique of "surveillance capitalism" highlights how these technologies, even when packaged as consumer conveniences, can create systems in which individuals feel perpetually monitored. The widespread push to "detox" from screens through designed retreats or device-free environments may offer a temporary reprieve but leaves the more pervasive digital transformations untouched.

In health-oriented projects, design trends like biophilic wards or collaborative clinics attempt to reduce stress and improve well-being. Yet these spaces only rarely engage with deeper shifts in how society conceptualizes health: from continuous digital monitoring to the normalization of rapid diagnostics and telehealth check-ins. Such solutions risk reaffirming an environment of perpetual self-surveillance without questioning how technology reshapes personal autonomy.

Thus, the core critique of our current defeating mechanisms is not that they lack ingenuity; it is that they often remain confined to addressing isolated challenges, none of which include our contemporary means of being colonized. As Henri Lefebvre (1991) suggests, altering physical space may temporarily mask deeper cultural and technological shifts that, when left unaddressed, continue to inform how we live and interact. These partial remedies can, at times, produce new forms of fragmentation—through oversight of how daily life, digital immersion, and human collaboration are braided together. The result is often a cycle of short-term successes followed by unmet needs as our cultural and technological realities evolve.

...ON DESIGN AND TECHNOLOGY

Integrating emerging technologies into built environmental design reflects the continuous interplay between technological advancements and increasingly complex human demands. On the one hand, new tools—data analytics, simulation platforms, AI, robotic environments, immersive realities, parametric modeling, among others—provide transformative possibilities for the conceptualization and realization of architectural and urban spaces. Concurrently, broad social changes, from remote working to the profound impact of social media, significantly reshape the ways people engage with their built environments. Despite the exciting potential these shifts suggest, architects, urban planners, and allied professionals consistently struggle to reconcile entrenched practices with the dynamic realities of contemporary society. Central among these challenges are:

1. limitations of current architectural systems,
2. the need for robust evidence-based approaches,
3. the limits of classical design programs in an increasingly digital world,
4. struggles in adopting new tools and systems,
5. the shortcomings of predictive analytics for human well-being, and
6. the unpredictability of social trends.

Collectively, these issues underscore the urgent imperative to recalibrate the interface between design, technology, and society—employing innovation without severing ties to historical, cultural, and emotional intelligences embedded in space. Architectural practice, historically codified through prescriptive regulations and formalist canons, remains largely constrained by linear workflows and static typologies (Lawson, 2006; Pérez-Gómez, 2016). Buildings are still too often conceived as inert, object-centric constructs rather than adaptable ecologies capable of mutating with sociotechnical flux (Till, 2009; Schneider & Till, 2007). This rigidity reflects not only regulatory inertia but also the curricular ossification of architectural education, where pedagogy remains centered on formal representation and tool-based novelties rather than ideological responsiveness (Webster, 2008; Salama, 2015).

Moreover, conventional architectural timelines—schematic design, design development, construction documentation, and construction administration—tend to isolate early design choices from real-world performance and user interaction post-occupancy. Architects rarely revisit completed projects years later to thoroughly examine actual usage and occupant satisfaction (Zeisel, 2006). Additionally, professional fragmentation persists, with architects handling form; engineers managing structural, mechanical, electrical, and plumbing systems; and sociologists or environmental psychologists, if involved at all, entering late and sporadically. Such compartmentalization restricts a comprehensive, interdisciplinary engagement with emerging technologies—like integrated sensor networks or interactive robotic systems—that inherently require multidisciplinary collaboration.

Addressing these limitations necessitates broader adoption of Evidence-Based Design (EBD), an approach bridging intuitive architectural decision-making with rigorous, empirically grounded inquiry. Initially popularized within healthcare architecture through Ulrich's seminal research illustrating how hospital views of nature improve patient recovery (Ulrich, 1984), EBD now permeates various sectors. Yet, hurdles persist in shifting architectural practices from anecdotal or so-called "best practice" decisions toward robust methodologies grounded firmly in empirical validation (Cooper Marcus & Barnes, 1999). Hamilton's framework identifies four progressive levels of EBD (Hamilton, 2003; Hamilton & Watkins, 2008):

1. Level 1: Designers rely on personal experience, informal observations, or industry norms.
2. Level 2: The design process draws from credible guidance—such as reputable journal articles, professional consensus, or pilot studies—but may not fully engage with robust research.
3. Level 3: The process integrates well-established findings, including peer-reviewed research with repeatable results, and carefully applies these insights.
4. Level 4: The design team contributes to the body of evidence through formal research, post-occupancy evaluations, and longitudinal studies, thereby advancing broader professional knowledge.

Many architects and engineers remain entrenched at Levels 1 or 2, partly because progressing to Levels 3 and 4 demands advanced methodological expertise, significant funding, and sustained interdisciplinary collaboration, rarely typical within conventional practice (Salonen et al., 2020). Additionally, clients often hesitate to invest resources in extensive occupant studies or post-occupancy evaluations, viewing them as overly costly or time-consuming. Pursuing deeper engagement with evidence-based methodologies can significantly enhance confidence in applying emerging technologies—like occupancy tracking, advanced air-quality sensors, and AI-driven simulations—thereby providing measurable validation of design strategies for stakeholders and end-users alike.

Evolving Classical Programs of Architectural Systems

Architectural programming, historically bound by typological determinism, continues to shape buildings as monofunctional containers defined by rigid spatial expectations. Yet, the dissolution of spatial certainties—catalyzed by the hybridization of public and private spheres through remote work, streaming economies, co-living infrastructures, and digital culture—renders these inherited models increasingly obsolete (Montgomery, 2013). Today's building typologies operate in a fluid field where the program is no longer prescriptive but performative. Despite these tectonic shifts, most architectural programming still reflects anachronistic design logic.

Limits of Adapting to New Tools and Systems

Technological instrumentation—Building Information Modeling (BIM), parametric modeling, generative algorithms, and immersive simulation—has introduced new epistemologies of design, but their integration remains uneven and often superficial. While these tools promise precision, interoperability, and iterative feedback, their application rarely transcends visual optimization or formal manipulation. There are practical limits to what can be modeled and simulated. Even to the extent that modeling and simulation are technically possible, they may entail unbearable financial or schedule burdens. The epistemic core of design practice remains rooted in static authorship rather than dynamic orchestration.

Architectural education is partly to blame. While digital proficiency is now a curriculum staple, it is too often divorced from theoretical or methodological inquiry, and the complexities and nuances of practice. Graduates are trained in tools, not systems thinking, and not how to address complex legal, code, and building science concerns. As a result, tool use competence coexists with conceptual and practical fragility. Conversely, the unchecked valorization of computational tools has led to their uncritical adoption. Algorithmic design, while seductive, often encodes unexamined biases and replicates prescriptive ideologies embedded in its training data (Sassen, 2017). Parametricism, for instance, purports to offer radical freedom yet often delivers formalist iterations devoid of sociocultural sensitivity.

The challenge, therefore, is not merely to adopt new tools, but to cultivate frameworks that critically interrogate their assumptions, affordances, and limits. Architecture must cease to be a passive recipient of technological innovation and instead become an active, reflective interlocutor in its development.

LIMITS OF PREDICTIVE ANALYTICS FOR HUMAN PERFORMANCE AND WELL-BEING

Predictive analytics has emerged as a cornerstone of performance-based design. Through the aggregation of behavioral data, climate metrics, and occupancy patterns, predictive models seek to preempt user needs, optimize comfort, and minimize inefficiencies. Yet, these promises are predicated on a dangerous assumption: that future conditions can be stabilized, coded, and forecasted.

This predictive rationality collapses when faced with sociocultural and building science complexities. Datasets—no matter how large—are only as representative as their sources. They often reflect static temporal slices, omit minority behaviors, entail many assumptions and limitations, and universalize culturally contingent norms (Montgomery, 2013). Moreover, behavioral prediction presumes rational occupancy, population homogeneity, ignoring the irrationalities, contradictions, and symbolic gestures that animate space.

Even to the extent that resources are dedicated to very large models and simulations, there are limits to predictability. As Taleb (2007) explains with respect to calculating the trajectory of a pool ball on a pool table,

> "*If you know a set of basic parameters concerning the ball at rest, can compute the resistance of the table (quite elementary), and can gauge the strength of the impact, then it is rather easy to predict what would happen at the first hit. The second impact becomes more complicated, but possible; you need to be more careful about your knowledge of the initial states, and more precision is called for. The problem is that to correctly compute the ninth impact, you need to take into account the gravitational pull of someone standing next to the table (modestly, Berry's computations use a weight of less than 150 pounds). And to compute the fifty-sixth impact, every single elementary particle of the universe needs to be present in your assumptions! An electron at the edge of the universe, separated from us by 10 billion light-years, must figure in the calculations, since it exerts a meaningful effect on the outcome*".

(Taleb, 2007, p. 178)

If it is this difficult to predict the behavior of a pool ball mere seconds into the future, then how accurate can a model of complex, Cyber-Physical Systems (CPS) ever be? Similarly, Trojanová et al. (2023) point out that there are practical limits to how much modeling, simulation, and prediction is possible given a dataset. It is always in everyone's interest to make models as parsimonious as possible.

Therefore, the reliance on analytics must be tempered by ethnographic insight, post-occupancy critique, and interpretive methods that privilege narrative, memory, and perception. Without this recalibration, environments utilizing predictive analytics risk scripting human life according to reductive metrics, reinforcing a culture of control rather than care.

THE UNPREDICTABILITY OF SOCIAL TRENDS

Societal behaviors do not evolve linearly. They leap, rupture, and reverse. The COVID-19 pandemic rendered decades of workspace standardization irrelevant in mere weeks, as the domestic sphere absorbed previously public functions. In parallel, global platforms modulate aesthetic preferences and spatial expectations at unprecedented velocity, creating trends that bypass architectural intent altogether (Sennett, 2013).

The volatility of cultural norms undermines master planning and long-term architectural programming. Static typologies—malls, airports, suburban enclaves—face rapid obsolescence under shifting patterns of work, leisure, and mobility. In this climate, architectural permanence becomes less a virtue than a liability.

Designers must therefore develop adaptive logics that privilege elasticity over monumentality. Adaptability is not simply a matter of movable walls or flexible leases; it is a philosophical stance that accepts incompletion as a condition of relevance. Pop-up typologies, reprogrammable public space, and provisional infrastructure are not ephemeral byproducts but essential architectural strategies in times of cultural instability (Jacobs, 1961; Folke et al., 2010). The challenge is to cultivate forms that do not merely endure but evolve—to build in ways that remain coherent precisely by refusing closure. It is in this ethos of calibrated openness, not predictive certainty, that contemporary architecture might begin to reflect the full complexity of its time.

ECOLOGICAL NICHE CONSTRUCTION

ENC reframes the long-standing evolutionary assumption that species passively adapt to pre-given environmental conditions. There are four mechanisms for evolutionary adaptation in the ENC model:

1. ecosystem engineering,
2. modification of selection pressures,
3. ecological inheritance, and
4. adaptation.

ENC foregrounds the organism as an active agent that modifies its surroundings in ways that recursively shape its own evolutionary trajectory (Odling-Smee et al., 2003; Laland & Sterelny, 2006). These modifications are structurally consequential—extending across generations via what has been termed ecological inheritance (Laland et al., 2001). In this light, organisms—human and non-human—no longer occupy space but construct it, embedding feedback mechanisms that mediate not only survival but long-term morphological, behavioral, and systemic shifts.

This theoretical framework holds particular resonance for architectural discourse, not in its ecological romanticism, but in its methodological implications. The built environment, traditionally regarded as inert infrastructure, is exposed as a recursive field of interventions—designs shaping inhabitants who in turn reshape their habitats. From dam-building beavers to urban humans, ENC challenges the architect-as-hero model: design is no longer an imposition but a participation in ongoing environmental recalibration.

ENC's broader relevance—anthropologically, biologically, and technologically—will be further explored later in this book. For now, it serves as a conceptual provocation. In doing so, we move from designing for the environment to designing with it, under conditions where environmental design is iterative and recursive, and agency is never singular.

THE LIMITS OF OUR HEROES...

Across this investigation into contemporary mechanisms of resistance, we have followed architectural, technological, and systemic responses that—despite surface-level sophistication—often disintegrate when asked to engage with deeper, structural entanglements. Today's design "heroes"—whether through adaptive infrastructures, decentralized systems, ecological simulations, or parametric modeling—appear less as disruptors and more as refined continuations of the very logics they claim to challenge. The result is not transformation, but repetition: recursive gestures of mitigation posing as acts of revolution.

What follows in Part III is not a continuation, but a rupture. Titled "*Sufferings of Humans in the 21st Century*," it moves from the mechanics of control to the lived consequences of their failure. Here, critique inhabits the aftermaths—digital spaces saturated with surveillance, infrastructures that neglect climatic realities, sterile healthcare facilities that alienate rather than heal. These are not theoretical failures; they are spatialized traumas.

Part III refuses prescription. It attends instead to atmospheres of suffering—those that emerge not from failure alone, but from design's misplaced certainties. Where Part II exposed the heroes and their limitations, Part III reveals what they've ignored: fragility, slowness, contradiction, grief. The path forward is about more than optimization or adaptability, but through radical presence—architectures that dwell in ambiguity, that listen rather than react, that resist becoming instruments of control that support the agents who occupy them.

Only in rejecting the urgency to resolve can we begin to notice the architectures of harm. The suffering of the present is not accidental. It is spatial, patterned, and designed.

Part III Sufferings of Humans in the 21st Century

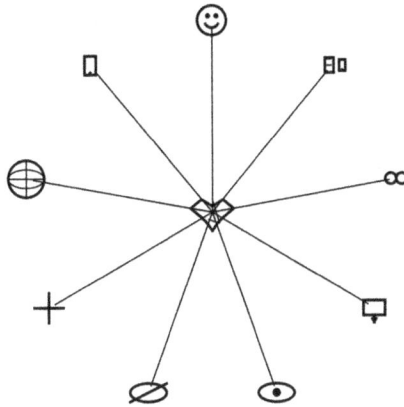

CONTEMPORARY SUFFERINGS

"...no beginning or end; it is always in the middle, between things, interbeing, intermezzo".

(Deleuze & Guattari, 1987, p. 25)

Navigating contemporary life involves continuous negotiations with socioeconomic disruptions and pervasive technological transformations. Echoing Deleuze and Guattari's (1987) concept of perpetual deterritorialization and subsequent reterritorialization, individuals are repeatedly displaced from familiar settings into fleeting new environments. This ceaseless movement generates instability, challenging human capacities to maintain psychological, physiological, and existential stability amid constant turbulence.

Our present era unfolds within accelerating economic transformations, social dislocations, and incessant technological developments—innovations touted as pathways to improved efficiency but often introducing significant psychological stressors and emotional fragmentation. From "smart cities" filled with algorithmically driven surveillance, to evolving modes of employment and communication, people frequently find themselves within environments that disregard genuine human vulnerabilities. Though presented as progressive, these changes expose the limitations of rigid architectural frameworks, incapable of adequately responding to crises or addressing traumas characteristic of our times, thus leaving occupants perpetually unsettled (Klingberg, 2009; Carr, 2011).

Such tensions culminate prominently in what Sherry Turkle (2011) defines as the "digital dilemma," wherein built spaces remain rigidly unresponsive to the pervasive intrusion of digital technologies. Privacy erosion and intensified surveillance become routine experiences as outdated infrastructure hastily accommodates monitoring equipment, eliminating any semblance of personal sanctuary. Individuals find themselves confined to public spaces—parks, libraries, cafés—bristling with sensors, cameras, and digital devices, with few opportunities for refuge. Architectural designs prioritize connectivity and device integration over genuine human interaction, reflected in inflexible table arrangements and regimented seating, thereby encouraging retreat into screens rather than fostering communal bonds. This inadvertent intensification of isolation leaves individuals both digitally connected and socially detached (Van Dijck, 2014).

Architectural inflexibility intensifies further in response to escalating environmental crises. Accelerated urbanization often neglects green space integration, intensifying environmental stress by depriving residents of restorative natural elements essential for psychological and physiological health (Kaplan & Kaplan, 1989). Dominated by concrete and steel, dense urban fabrics fail to swiftly accommodate necessary retrofits, leaving communities vulnerable and incapable of effective adaptation to rapidly shifting ecological conditions (Folke et al., 2010).

Similarly, architectural rigidity extends into healthcare infrastructure, giving rise to a paradoxical phenomenon where sterile hospital environments intensify rather than alleviate patient distress (Sapolsky, 2004). Clinical spaces persist with austere, rigid layouts, restricting daylight access, personalization, and family engagement—factors crucial to patient well-being. Moreover, urban design rarely emphasizes preventative care strategies such as wellness hubs or meditation gardens. Instead, stringent zoning laws, regulatory frameworks, and budget constraints confine such spaces to peripheral or exclusive areas, maintaining a reactive rather than proactive health system paradigm (Mumford, 1938).

The architectural failure to address the escalating pressures of the contemporary "economy of attention" also warrants attention. Open-plan office designs, once symbols of collaborative innovation, now exacerbate employee stress by overloading workers with incessant auditory and visual stimuli (Postman, 1985). Static workstations within standardized grid layouts preclude flexibility, impeding concentrated thought and psychological tranquility.

Simultaneously, rapid societal shifts toward remote work and gig economies highlight architectural inflexibility's role in perpetuating economic disparities (Scharff & Dusek, 2003). Social fragmentation and polarization also manifest architecturally through designs geared toward exclusivity, cementing divisions between disparate communities (Habermas, 1989). Public spaces intended for inclusive use—libraries, cultural centers, community halls—often remain unadaptable to multicultural or diverse community events, perpetuating exclusion through architectural and social inflexibility (Polanyi, 1944; Lyon, 2018).

Amid interlocking crises—digital surveillance, environmental fragility, social isolation, and economic uncertainty—architecture ought to serve as a resilient, adaptive infrastructure responding dynamically to human needs. However, prevailing architectural practices, dominated by inflexible, single-purpose designs, intensify rather than alleviate contemporary anxieties (Feenberg, 1999).

Increasingly, scholars and practitioners advocate multifunctional, adaptable architectural strategies prioritizing human-centered design, natural integration, and reduced digital intrusion (Kitchin, 2014). Biophilic principles, incorporating greenery and natural elements, aim to bolster psychological resilience (Kaplan & Kaplan, 1989), while flexible spatial arrangements using modular furnishings and movable partitions offer potential responsiveness to fluctuating circumstances (Norman, 2013).

Ultimately, addressing the complex sufferings of contemporary life demands architectural approaches emphasizing genuine adaptability, responsiveness, and holistic human well-being. Instead of forcing individuals into rigidly structured environments, buildings must dynamically accommodate shifting technological, ecological, health-related, and societal conditions (Odling-Smee et al., 2003). Recognizing architecture's role in either exacerbating or alleviating modern traumas could catalyze a fundamental shift from reactive design practices toward proactive, human-centered models.

In subsequent pages, this analysis explores how architectural inflexibility exacerbates contemporary challenges, beginning with the digital dilemma—examining surveillance and alienation within digitally infiltrated public spaces. Following this, the discussion highlights climate-induced vulnerabilities within unresponsive environmental designs, before addressing healthcare paradoxes accentuated by sterile hospital environments.

HUMANS AND THE DIGITAL DILEMMA...

It begins on a rain-soaked morning, in a metropolis perpetually awake, where neon reflections of corporate insignias shimmer across wet concrete. Glass façades and stark steel frameworks envelop your every movement, structures engineered to embody efficiency, rationality, and futurism. Yet an unsettling discomfort persists, subtly lurking at the periphery of your awareness. Overhead cameras and infrared sensors observe silently—mechanical eyes tracing each step, reminiscent of a shadowy digital thriller. You linger within a public atrium purportedly designed to foster urban sociability; instead, it hums with an invisible exchange of data streams, capturing your expressions, movements, even emotional states. No nook remains untouched by sensors adept at decoding your private impulses. Architecture itself becomes complicit, ensnared within a rigid geometric order that offers no sanctuary or reprieve.

Surrounding screens pulse continuously with tailored content—unstoppable feeds of targeted advertisements, geo-based notifications, and persistent social media alerts. Embedded deeply within walls lie unyielding infrastructures of cameras, motion detectors, and thermal sensors. Urban planners once celebrated these designs as "smart," yet navigating these spaces reveals "smart" as merely a softer term for constant oversight—perpetual data acquisition promising convenience while simultaneously corroding personal privacy (Zuboff, 2019; Kitchin, 2014). Within these cityscapes, architecture has ceased to merely shelter; it now surveils.

Yet this trajectory was never predetermined. Early advocates emphasized adaptability, urging that architectural designs should flexibly integrate emerging technologies without eroding individual privacy. Such calls were eclipsed by promises of heightened efficiency, cost-effective construction models, and the irresistible lure of uninterrupted connectivity (Sadowski & Bendor, 2019; Morozov & Bria, 2018). Consequently, urban environments now embody an inflexibility unable to adjust adequately to rapidly evolving digital paradigms, intensifying human distress through psychological burdens induced by constant observation, documentation, and data-driven scrutiny. We are building the instruments of our own captivity. This pervasive issue engenders what is termed an "Unadaptability Layer": surveillance systems, once integrated, resist recalibration or removal (Sadowski, 2020). Walls conceal intricate electronic command centers; corridors house discreet drone recharging ports; lobby ceilings bristle subtly with wide-angle surveillance lenses. Interior spaces remain locked into singular functions, affording no flexibility to occupants attempting to reclaim privacy. Family residences find themselves incapable of deactivating embedded "energy-monitoring" cameras without significant structural overhaul. Similarly, workplaces embedded with analytics software feed continuous occupant data to third-party commercial entities, transforming physical environments into silent informants (Moore & Robinson, 2016).

This digital entanglement is hardly limited to corporate office towers or luxurious retail spaces. Even civic plazas, originally envisioned as spaces for communal gatherings, incorporate sensor networks meticulously tracing your trajectories, routines, and inferred desires (Couldry & Mejias, 2019; Lyon, 2018). Architects failed to include an "off switch" for ubiquitous watchers perched upon façades and rooftops. Thus, urban environments mutate into halls of digital mirrors, constantly reflecting your actions within distant databases. Discussions around "smart city" innovations often occur with uncritical enthusiasm, yet lived realities frequently resemble invasive data streams populating corporate and governmental analytics dashboards, shaping consumer identities, determining risk categories, and even influencing social credit systems.

SCENARIO 1: PRIVACY EROSION AND SURVEILLANCE OVERLOAD

Paulette steps from the neon-bathed streets into the imposing shadow of Arco Silente, a shimmering tower whose glass and steel surfaces exude an unsettling omnipresence. In the dim evening glow, she detects a silent intelligence flowing through concealed vents, an unseen watcher murmuring wordless demands for the surrender of her privacy the instant she crosses its threshold.

Meeting her friends—Jules, Camila, and Hafiz—in the expansive lobby, Paulette glimpses anxious reflections mirrored on the pristine floor tiles. They cannot evade the vigilant gaze of Arco Silente; sensors above pivot at their arrival, absorbing data on temperature, heartbeats, even minute emotional cues. Jules nervously fidgets with a small signal disruptor he assembled, hoping it might jam surveillance feeds. Within seconds, it fails.

"I have you mapped," the building seems to intone, though it possesses no voice. Paulette imagines its unspoken assertions in every automated door and subtly glowing LED indicator. Camila gestures toward an innocuous orb discreetly nestled within the ceiling. "They claimed it's merely for health monitoring," she murmurs uneasily, "but it's scanning far beyond body temperature. It knows exactly who we are, everything we do."

Hafiz attempts optimism, yet the oppressive presence of the structure envelops them, suffusing the air with static tension. "Is this truly making us safer," he asks hesitantly, "or is it surveillance merely for surveillance's sake?" Arco Silente does not reply directly; only the mechanical murmur from its ventilation system whispers a faint response.

As they wander the open-plan corridors bathed in meticulously curated illumination, Paulette notes the irony—this once-lauded example of architectural futurism has become a rigid sentinel, its cameras eyes, its hallways conduits of relentless data extraction. She searches for some quiet niche, a pocket of privacy hidden from digital prying, but discovers no refuge remains. An almost palpable satisfaction emanates from the architecture, smugly aware that it has them thoroughly monitored.

"I tried adding privacy screens," Camila sighs, recalling earlier resistance. "Building administrators dismissed it, citing 'system integrity' concerns." Jules nods gravely. "Remove even one sensor and the whole system collapses into error messages." "If you prevent or block sensors, then appliances lights, HVAC, & audio-visual systems cease to work until they can sense you again. There is no offline mode for using the apartment's systems."

Overhead screens flash updated occupant metrics, oblivious to their mounting unease. SECURE AND EFFICIENT, announces the digital message. Paulette nervously rubs her arms, aware even this instinctive gesture is now archived and analyzed to determine if it portends changes to circulation, organ function, hormones, or psychological states, and in response calling up targeted ads to address the perceived issues. "All I want is one private moment," she whispers, barely audible.

They retreat outdoors, stepping onto rain-slick pavement, turning to view the structure radiating an unyielding sheen under streetlights. "We've become mere data points," Jules mutters with quiet bitterness. "Nothing will shift until someone tears the building's systems from its core."

Paulette hesitates, feeling Arco Silente's metallic silence pressing upon her, an ever-present reminder of how architecture can transform into a captor, leaving no room for spontaneous, unobserved life.

The relationship between digital infrastructures and physical architecture reveals a deep contestation at the intersection of code and concrete, an *interface where built spaces lock in technological imperatives. Once develop- ers finalize plans and officials inaugurate these buildings, occupants find their fates sealed. Each structure embodies the design priorities of its inception; where maximum surveillance became paramount, buildings stand as imposing monuments of technological intrusion. Unlike a smartphone, smart watch, AR headset, or tablet, it is not possible to just drain the apartment battery, or turn it off, and throw it in a drawer.*

Surveillance devices, once installed, become permanent fixtures, like mecha- tronic guards perched on every cornice. Occupants must adapt or escape. Entire urban districts, developed under the premise that increased data extrac- tion benefits society, often entrench structural ideologies that leave little space for retreat, respite, or genuine human intimacy. The pursuit of efficiency and data-driven governance in smart cities often leads to environments where sur- veillance is pervasive, and personal freedoms are curtailed.

SCENARIO 2: SOCIAL MEDIA-DRIVEN ISOLATION IN PUBLIC SPACES

Paulette emerges from the subway onto a plaza dominated by Cortexus, an immense arrangement of sleek concrete wings and neon terraces designed explicitly to magnetize smartphone signals. She arranged this rendezvous with friends—Nida, Damon, and Aisha—to find relief from daily pressures, yet upon stepping onto Cortexus' meticulously engineered paving stones, she senses the architecture's invisible pull toward digital engagement at the expense of meaning- ful human connection. Subtle electronic currents pulse in the air; benches, tables, and raised platforms bristle with charging stations and reflective digital screens.

Spotting Damon first, Paulette sees him hunched at an immovable, wire- framed stool in a café row rigidly aligned to face flashing holographic news streams rather than fellow patrons. "Hey," Damon calls, half-heartedly waving a tablet. The chair beside him similarly cannot be moved, positioned deliber- ately so occupants remain fixed upon bright advertisements or outward-facing data projections. Paulette smiles faintly, noting Damon's distracted gaze shift- ing continually between her and flickering screens at his peripheral vision.

Nida arrives breathlessly, gesturing around Cortexus. "A review praised this place as the pinnacle of digital plazas," she remarks. "Constant WiFi, aug- mented reality layers—it's supposedly an informational utopia." Yet rigid seat- ing layouts and constricted pathways hinder interactions among groups larger than pairs. "It's isolating," she concedes softly, voicing Paulette's unspoken discomfort.

Approaching, Aisha pushes past an awkwardly placed half-sphere stool. "You can't rearrange any seats," she murmurs, nudging the stool's immovable base.

"All bolted down so you're forced to watch big screens or your own device."
Her laugh carries quiet resignation. "No chance for real conversation here."

Behind Cortexus, towering digital billboards display continuous social media updates. Visitors stand scattered, absorbed by their individual screens, even in groups appearing fragmented, each responding solely to personal notifications rather than engaging authentically with companions. Paulette observes how intricate pathways funnel visitors toward carefully arranged selfie spots, absent communal seating—design strategies documented to encourage shallow interaction that is maximally monetized rather than deeper social bonds.

From concealed speakers, Cortexus' artificial voice greets them: "Enjoy free connectivity. Tag us in your uploads." The message blurs greeting and instruction. Damon sighs openly, musing on how these environments, though never meant to feel so distant, ended up driving humans apart, reducing contact to digital superficialities rather than meaningful dialogue. Nida gestures toward a nearby "communal hub" advertised prominently—only to discover it consists solely of clustered charging points lacking communal tables. "The device matters more than the person," Aisha remarks quietly.

Together, they linger in a plaza structured for endless virtual engagement yet starved of tangible human warmth, unless they pay a lot of money. Paulette taps an empty seat beside her. "Let's make our own circle," she proposes wryly. Even within Cortexus' isolating geometry, they strive briefly to defy the architectural imperative that insists on solitude, attempting, if only momentarily, to reclaim genuine connection (without having to upgrade to it) in an environment engineered to fragment and distance.

HUMANS AND ENVIRONMENTAL CRISES...PHYSICAL AND MENTAL HEALTH

Modern urban landscapes increasingly confront environmental volatility, driven by escalating climate disruptions, resource scarcity, and erratic weather patterns. Architecture historically aligned with stable seasonal and regional climates now faces unprecedented pressures—extreme heatwaves, sudden flooding, irregular temperature fluctuations, and declining air quality—that amplify physical and mental distress for inhabitants. Structures initially adequate under predictable conditions frequently intensify human suffering amid climatic unpredictability.

The built environment can either mitigate or exacerbate these impacts. Buildings equipped with dynamic ventilation or adaptive insulation demonstrably alleviate occupant discomfort during temperature extremes. However, many contemporary urban constructions remain dependent on obsolete materials and rigid infrastructures, inadequately prepared for rapid climatic shifts. Such architectural rigidity intensifies thermal stress, leading to health complications, disrupted sleep patterns, and increased incidences of anxiety and depression (Environmental Health Perspectives, 2024).

Further, urbanization reduces natural buffers that traditionally moderated urban environments. As cities expand, wetlands and forests that absorbed stormwater and regulated local temperatures are replaced by impermeable surfaces, exacerbating flooding and amplifying the urban heat island effect. Conversely, urban green infrastructure—such as rooftop gardens, tree-lined streets, and urban forests—demonstrably mitigates these extremes by reducing urban temperatures and improving air quality.

Mental health equally suffers within sterile, inflexible architectural spaces devoid of biophilic elements. Extensive research confirms that exposure to natural features significantly reduces stress hormones like cortisol, enhances mood, and boosts cognitive functioning. Residents consequently find themselves enveloped in architecture incapable of responding to environmental variability, intensifying anxiety—not solely from climatic threats but also due to structures' incapacity to offer protective buffers. Work environments exacerbate discomfort through inflexible heating, ventilation, and air-conditioning systems that ignore occupant-specific comfort or preferences (Environmental Health Perspectives, 2024).

Climate-driven psychological stressors compound these issues further. Persistent anxiety linked to imminent storms or extreme heat—a condition increasingly recognized as "climate anxiety"—exacerbates psychological strain (Environmental Health Perspectives, 2024). Architectural inflexibility, exemplified by the absence of adjustable shading or resilient barriers, magnifies such anxieties by emphasizing occupant vulnerability.

Cities lacking adequate green spaces experience heightened stress, diminished cardiovascular health, and weakened social cohesion—collectively described as conditions of "concrete suffocation." Urban planners frequently undervalue green infrastructure, categorizing it as a non-essential luxury, thereby complicating its integration post-construction. Prevailing development paradigms—focused on immediate economic returns rather than long-term ecological resilience—often hinder such cross-disciplinary cooperation.

As climate volatility intensifies, the consequences of architectural inertia become starkly visible. Heatwaves strain power grids, floods overwhelm urban drainage systems, and severe weather exposes inadequately reinforced buildings. Occupants face amplified physical hardship, property losses, and chronic anxiety as their built surroundings fail to provide adequate protection. Hence, urban environments shift from safe havens to additional sources of distress and trauma.

Socioeconomic disparities further compound these vulnerabilities. Communities residing within aging and inadequately maintained buildings disproportionately experience climate-induced stress, reporting higher incidences of related mental health conditions (Environmental Health Perspectives, 2024). Responsive architectural modules that adjust to seasonal or climatic variability offer occupants autonomy, effectively reducing climate-related anxiety. For instance, sensor-activated glazing that dynamically adjusts opacity during intense sunlight stabilizes indoor temperatures and mitigates occupant stress, illustrating architecture's potential evolution from static structures into active facilitators of human health.

Ultimately, the escalating environmental volatility of the 21st century profoundly impacts human health, both physically and psychologically. Architectural rigidity—characterized by obsolete materials, fixed designs, and inadequate green integration—intensifies these stresses. As climate unpredictability collides with inflexible urban constructions, occupant anxiety and vulnerability rise significantly. Consequently, urban spaces devoid of adaptive and ecological strategies continue exacerbating mental and physical strains. A critical imperative emerges for architecture to transcend rigid frameworks, embracing fluid, ecologically-informed designs that actively enhance human resilience, transforming built environments into genuine sanctuaries amid ecological uncertainty.

SCENARIO 3: CLIMATE-INDUCED INFLEXIBILITY

Paulette steps off the maglev tram into a district named Aeroveil, a futuristic sprawl of high-rise blocks rumored to be an "urban marvel." The city's overhead broadcast had boasted of Aeroveil's grand boulevards and glittering façades. But as a stifling heat wave grips the region—an unexpected, record-breaking spike—Aeroveil's glass towers seem to radiate scorching air back onto the streets. Pavement and walls alike reflect a punishing glare under the midday sun.

Her friend Jericho waits by a sprawling archway that leads into an atrium. He's leaning against steel panels that are too hot to touch. "They said these materials were revolutionary," Jericho mutters, wiping sweat from his forehead. "Now they turn the place into a furnace." Together, they enter the atrium, waves of heat following them in. Paulette feels the dryness in her throat, a pain that reminds her of how unprepared this architecture is for climate extremes.

Inside, the building's central air system roars feebly. The overhead vents blast warm air instead of cool, locked by a preset schedule. Aeroveil was never designed for such sudden and brutal temperature spikes; the control algorithms are archaic, set to moderate mild fluctuations at best. The plasmatic glow of digital panels, normally a signature of futuristic design, compounds the heat by adding extra lumens and device-generated warmth.

> *"They promised me adaptive cooling," Jericho laments, pointing at a brochure pinned to a wall. "Look at that: 'Integrated thermal management for occupant comfort.' Lies!".*

They move deeper into Aeroveil, passing an enclosed courtyard that might have been pleasant in moderate weather. Now it bakes under the glass roof—greenery withered, the fountain nearly evaporated. Residents meander with a pained slowness, many carrying portable fans or cold water packs. A sense of desperation is palpable. The temperature outside soared by another ten degrees earlier, but the building can't compensate quickly enough.

> *"I studied the data last week," Jericho says, referencing an article by Santamouris (2015) pinned in his phone. Heat mortality soared in older districts with rigid construction. This place is new, but the design is still stuck in the past.*

By late afternoon, a local weather alert warns of rolling blackouts—excessive AC usage has strained the grid. Paulette glances at her watch, noticing her own pulse racing. She's read about how climate-induced stress intersects with rigid architecture to provoke anxiety and insomnia. Now she feels it first-hand: the walls around her no longer shelter but trap.

Workers attempt a patch: hooking up portable air units in the corridor. But Aeroveil rejects the ad hoc wiring, its built-in safety protocols blocking modifications. One official mutters about "maintaining system integrity," ignoring the pleas of panting residents. Jericho curses under his breath, stepping away to find some cooler corner—only to discover none exist.

> *"We could open windows," Paulette suggests, half-joking. She eyes the sealed glass panels overhead. "But they're not real windows. They're just illusions of openness, all fixed."*

At dusk, they join a crowd near an emergency kiosk. People gather in stifling clumps, searching for any official instructions. Heat radiates off the building's metal exoskeleton, the outside temperature barely dropping. A manager confesses that even if they wanted to swap out the walls for reflective surfaces or install vents, they can't—Aeroveil was built in a way that prevents modification, its materials locked in place.

> *Paulette and Jericho slip back into the swirling city night, sweat laced with frustration. They glance up at the tower's shimmering geometry, once sold as a beacon of modern living. In the face of this unprecedented heat wave, the tower stands inert, unable to adapt, turning from a proud edifice into an oppressive shell that intensifies the very climate crisis it was meant to endure.*

SCENARIO 4: ABSENCE OF GREEN SPACES IN DENSE URBAN AREAS

Another day, another part of the city: Paulette meets her colleague Rayen at Beton Crossing, an urban labyrinth of narrow corridors and colossal apartment blocks. The façade is pure concrete, with faint traces of shallow designs that once aimed to seem "artistic." In the haze of midday sun, the entire district feels like a compressed vault.

Rayen points to a mural of painted vines on a building wall. "This is the closest thing we have to greenery," she jokes. They slip between walkways that swerve unpredictably, overshadowed by the bulk of the architecture rising overhead. It's a place of no real vegetation—only illusions painted onto the concrete.

> *"Ever wonder why we have no parks?" Rayen muses. "They maxed out the land for apartments and offices, and now there's nowhere to retrofit green space."*

Paulette notices the tension in her shoulders as they traverse these dull corridors. She recalls studies proving that even small urban gardens or pockets of trees can alleviate stress (Lee & Maheswaran, 2011). *But in Beton Crossing, planters or rooftop gardens are nonexistent; the architecture left zero flexibility for natural enclaves. Residents talk about headaches, insomnia, and a sense of claustrophobia—symptoms of what some environmental psychologists call "concrete suffocation"* (Bratman et al., 2015).

At midday, they pass a courtyard rumored to be "open." In truth, it's a loading bay overshadowed by walls so high that sunlight barely touches the floor. A single sapling stands in a pot near a trash receptacle. Its leaves are scorched from lack of water. Rayen steps closer, trying to feed it a bit from her water bottle. Yet it's clear the building's drainage system doesn't support plant life.

> *"I asked city planners about turning this area into a micro-park," Rayen says. "But they said the building's foundation can't handle the root systems. They built everything so rigidly that we can't add real soil or infiltration basins."*

At a nearby café, metal benches are anchored in uniform rows, each facing a large digital wall. People sit alone, headphones plugged in. No trees, no canopies, not a single stretch of grass. The so-called "public realm" is a sliver of concrete guarded by tall fences. Many linger only briefly, as if the environment itself discourages lingering.

A wave of sirens echoes down a corridor. Water main break, someone says— another sign that Beton Crossing can't handle the strain of daily life. The thick, unyielding surfaces above ground reflect heat, while the subterranean pipes degrade from overuse. The city scrambled to find a solution, but with no open soil to absorb overflow and no green corridors to moderate the microclimate, flooding and water damage are a constant threat (Fletcher et al. 2015).

Paulette and Rayen duck into an alcove for shade. Sweat trickles down Paulette's neck. She thinks of the balanced squares or communal gardens she's glimpsed in old photographs. Here, the building's designers prized maximum density and forgot the simplest relief: nature.

> *"If we want greenery, we'd have to demolish half the block," Rayen murmurs. "These walls can't be moved or reconfigured. It's all or nothing."*

As they continue on, the edges of Beton Crossing reveal more looming towers, culminating in a skyline of solid, gray blocks. The monotony weighs on Paulette—stirring an itch for open skies, for color, for leaves rustling in a breeze. She recalls medical research linking a lack of green spaces to heightened anxiety and social isolation (Evans, 2003). *It's not just about aesthetics, but mental survival.*

They eventually emerge onto a busy thoroughfare, relieved to see a single row of ornamental shrubs near a bus stop. The shrubs are meager, but to

Paulette, they might as well be an oasis. She stands there a moment, letting her eyes rest on something alive. She can almost feel her pulse steady. Then the city crowd surges past, reminding her that real green spaces remain absent from this rigid sprawl.

"There's a tension here," she says softly to Rayen, "between how we want to live and how the architecture confines us."

Rayen nods, glancing over the towering horizon of concrete. A breeze stirs the meager shrubs, a small gesture of life amid the gray. And in that fleeting instant, Paulette hopes for a future where walls might be coaxed to recede, letting in roots, leaves, and an enduring sense of calm that the city so desperately craves.

THE HEALTH PARADOX...

Contemporary healthcare architecture often strives for innovation—embracing cutting-edge technology, rapid interventions, and streamlined efficiency. Yet, despite intentions of healing, many clinical environments paradoxically intensify the distress they aim to alleviate (Ulrich et al., 2004). Rigid hallways, monochromatic palettes, and unyielding surfaces common to hospital complexes reflect industrial-era priorities, prioritizing functional throughput over holistic patient wellness (Cooper Marcus & Sachs, 2013). Entrenched design traditions, ensuring operational hygiene and systematic care, paradoxically perpetuate environments undervaluing psychological wellness, intensifying patient anxiety, and potentially prolonging recovery (Huisman et al., 2012).

Historically, hospital architecture emphasizes sterility and infection control—requirements undeniably crucial for patient safety (Malkin, 1992). Yet, excessive utilitarian spaces lacking comforting aesthetics or personal elements foster isolation and heighten patient anxiety, negatively affecting resilience (Joseph & Ulrich, 2007). While sterility is vital, overly minimalistic designs often overshadow therapeutic virtues such as natural illumination, inviting textures, and personalized environments (Cooper Marcus & Sachs, 2013). Extended confinement in stark clinical settings leads to heightened discomfort, stress, and emotional alienation (Ulrich et al., 2004). Efforts integrating natural components—gardens, views, personalized aesthetics—are frequently demoted due to infection protocols and budgetary constraints (Huisman et al., 2012).

Medical technology advances unintentionally amplify impersonal hospital atmospheres. Persistent beeps, glowing screens displaying vital statistics, and the subtle hiss of intravenous pumps under harsh lighting echo dystopian fiction (Cooper Marcus & Sachs, 2013). Collectively, these instruments overwhelm patients, converting rooms into mechanical arenas where identity feels secondary to clinical efficiency (Joseph & Ulrich, 2007). Vulnerable patients seeking emotional refuge find themselves engulfed in artificial alarms and sterile whiteness, deepening their sense of alienation (Ulrich et al., 2004).

Holistic health—encompassing emotional comfort, spiritual solace, and community support—requires adaptable, nurturing environments (Joseph & Rashid, 2007). Research consistently demonstrates how natural elements—garden courtyards, plant-filled atriums, even virtual landscapes—can lower stress, alleviate pain, and accelerate recovery (Ulrich et al., 2004). Yet hospitals minimally incorporate these, often restricted to superficial additions like small healing gardens or houseplants (Cooper Marcus & Sachs, 2013).

This architectural rigidity mirrors societal tendencies favoring reactive over preventative healthcare interventions (Giles-Corti et al., 2016). Health services, mental health clinics, and fitness centers often reside on urban peripheries, perpetuating a crisis-driven model rather than proactive wellness strategies (Maas et al., 2006). Conversely, preventive measures—community exercise spaces, wellness hubs, counseling centers—are frequently excluded from urban frameworks, viewed as luxuries rather than essential infrastructure (Giles-Corti et al., 2016).

The COVID-19 pandemic vividly exposed inflexible healthcare infrastructure deficiencies, hospitals overwhelmed during patient surges, corridors lined with temporary beds, communal spaces abruptly transformed into wards, and staff pushed to exhaustion (Ranney et al., 2020). Architectural inflexibility exacerbates physical strain and emotional fatigue, intensifying stress for patients and healthcare professionals alike (Fenn & Wilber, 2022). Moreover, the shortage of preventive spaces intensifies this paradox (Maas et al., 2006). Dense urban zoning typically prioritizes commercial or residential development over integrated wellness facilities, limiting access and fostering sedentary lifestyles, delaying proactive wellness interventions (Giles-Corti et al., 2016).

Mental health advocates compellingly argue for embedding preventive wellness spaces—gyms, counseling pods, stress-relief areas—within residential and commercial contexts, reducing healthcare costs and enhancing community wellness (Giles-Corti et al., 2016). Yet, implementing flexible, multi-use spaces demands revising entrenched architectural codes and planning norms, often encountering resistance due to financial concerns or disruptions to traditional property values (Fisk, 2019).

Addressing the prevailing health paradox demands transformative shifts. Architects and urban planners must embrace EBD's methods and merge infection control with human-centric aesthetics (Joseph & Rashid, 2007). Modular patient rooms with adjustable lighting, customizable décor, and integration with natural elements represent necessary innovations. Cities must establish flexible, multi-use health hubs convertible among gyms, counseling centers, community kitchens, and clinics. Ignoring this critical inquiry risks intensifying patient suffering, converting healthcare facilities into anxiety-inducing zones rather than genuine therapeutic refuges (Fenn & Wilber, 2022). Thus, future healthcare design must prioritize adaptability, personalization, and preventive approaches, transforming rigid corridors into vibrant healing environments genuinely supporting human resilience.

SCENARIO 5: THE STERILE HOSPITAL ROOM AS A STRESSOR

Paulette enters the Stellaris Wing, a hospital extension acclaimed for its "futuristic design." The corridors, illuminated by LED lights, exude a stark pallor, with glossy tiles reflecting an institutional glare. Automated doors hiss open, revealing a series of identical patient rooms. Inside each, she observes sterile, whitewashed walls interrupted only by the muted beeping of medical monitors.

In Room 506, her friend Maya reclines against stiff pillows. The room is immaculate yet impersonal, devoid of personal photos—only a functional bed, side table, and a sealed window for infection control. Maya's shoulders slump under the weight of an environment that erases individuality. "I've only been here three days," she murmurs, "and I feel more like a specimen than a person."

A nurse enters, scanning Maya's vitals with a handheld device. The beep buzz of machines nearly drowns out the nurse's gentle greetings. Paulette recalls research indicating that monotonous hospital rooms can elevate stress hormones, potentially prolonging patient stays. She notices the barren walls and rigid furnishings—no curtains to soften the harsh lighting, no space for personal items.

> *"I brought you a poster," Paulette offers, unfurling a small print of a seaside view. She tries to tape it to the wall. Immediately, a sensor by the ceiling flashes red, a pre-recorded voice stating: 'Unauthorized modifications to the patient environment are not permitted.'*

Maya sighs. "Even a little color is disallowed. They said it might gather bacteria or compromise the cleaning protocol."

Her vitals beep on the overhead screen, lines charting her pulse as if mirroring her frustration. A locked cabinet filled with sterilizing gear dominates the room, overshadowing any attempt at personal comfort. The environment is rigid—benches fixed, the bed adjustable only to predefined angles. It's as though the architecture has declared war on spontaneity.

> *"It's suffocating," Maya whispers. "I can't crack a window for fresh air, can't adjust the overhead lights, and the walls stare back at me like they're made of ice."*

Paulette envisions the architecture as an unyielding fortress, whispering in mechanical monotone: I was built to keep you clean, not comfortable. She senses the environment draining Maya's hope, bridging the gap between medical necessity and emotional deprivation.

That evening, during a staff shift change, Paulette strolls through the main atrium. Generic motivational posters ("HEALTH = HOPE") line the corridor, ironically lifeless in such a rigid setting. She recalls studies suggesting that an overemphasis on sterility can overshadow the humanity of healthcare environments. A cleaning drone glides past, spraying antiseptic in a sweeping arc, leaving a cold chemical scent. Everything is safe, sanitized, and heartbreakingly impersonal.

Upon returning to Maya's room, Paulette notices her friend's shallow breathing. The monitors display stable vitals, but Maya's expression betrays the stress of emotional confinement. "I'd heal faster," she says softly, "if I could just feel alive here. Instead, it's like these walls drain color from my soul."

> *"I'll talk to the admin," Paulette replies, though she suspects they can't override the hospital's standardized blueprint. The architecture stands, unyielding: a minimal complex designed for infection control, but heedless of the deeper healing that patients crave. She lingers a moment, imagining how a single patch of personalization—a painting, a potted plant, or even a small family photo—might break the sterile tension, offering a glimmer of humanity in an otherwise mechanical domain.*

SCENARIO 6: LACK OF PREVENTATIVE SPACES IN URBAN AREAS

A week later, Paulette navigates the city to meet her colleague Jasper outside Harborview Heights, a cluster of towering apartments famed for their panoramic seafront vistas. She anticipates breezy communal terraces and invigorating open-air walkways. Instead, she finds tiered balconies locked behind tinted glass, each stacked above the other like sealed compartments. The sidewalk is a narrow strip flanked by heavy traffic on one side and a sterile building façade on the other.

Jasper gestures down the street. "Do you see a gym or wellness center anywhere? Because I sure don't." He scrolls through an urban zoning map on his phone, referencing studies that highlight the poor integration of health-promoting infrastructure in high-density residential zones. The entire district is designated for residential towers, leaving no legal allowance for exercise or mental health spaces.

They explore the ground level, where a small sign reads Community Health Services—Basement L3. Inside, they discover a dimly lit corridor leading to a cramped, windowless room. A single treadmill whirs feebly under flickering fluorescent bulbs. A volunteer at a desk apologizes for the limited hours. "We don't have funding for more equipment," she admits, referencing an architecture that never accommodated wellness. "They can't expand upwards or sideways. The building's design is locked."

> *"I tried to get a meditation corner installed in our lobby," Jasper muses, "but management said it would 'interrupt the building's traffic flow.' The entire concept of prevention gets relegated to leftover spaces. Nobody invests in real, multi-purpose rooms that can shift from yoga classes to therapy sessions to community workshops."*

They walk outside again, spotting a billboard that touts the city's latest hospital expansion. "Funny how they'll invest in big medical centers," Paulette remarks, "but not in places that stop people from getting sick to begin with." She recalls research indicating that green living environments significantly correlate with reduced morbidity rates.

"People end up in those sterile hospital rooms," Jasper adds bitterly, "because the city never gave them an easier path to daily well-being." They pass an elderly resident shuffling along the sidewalk, nowhere to sit or rest except a concrete planter rim. A small palm tree stands near-dead in the scorching sun.

As they turn a corner, Paulette and Jasper encounter a former community center, now boarded and eclipsed by fresh construction. A glossy sign proclaims the arrival of luxury condos boasting "state-of-the-art security," yet remains silent on wellness amenities. "We're trapped in a reactionary model," Paulette observes, recalling Perdue et al. (2003), who emphasized how zoning rigidity and planning inertia consistently undermine the integration of proactive, health-focused architecture.

Their final stop leads them to a neglected rooftop, reputedly offering a worthwhile view. Upon reaching it, they find a barren expanse—dust-covered concrete littered with disused antennas. The city sprawls beneath them, shimmering with layered highways and gleaming towers. "This could've been a rooftop garden or a yoga deck," Jasper says softly, running his hand along the cracked surface. "But the building codes never called for it. So it's empty concrete."

"Seems we're all locked out of prevention," Paulette murmurs, scanning the horizon. She spots healthcare complexes in the distance, each building an imposing fortress. "Hospitals loom tall, but the real help we need—spaces to stay healthy—is nowhere to be found."

As the sun dips behind the dense assembly of towers, the city's glow deepens. In that half-light, Paulette and Jasper perceive how decisively the built environment shapes daily life. Streets hostile to pedestrians, structures indifferent to physical activity, and urban designs relegating mental health resources to obscurity—all conspire to frame wellness as a luxury. They linger thoughtfully, envisioning a city reimagined with flexible zoning, adaptable communal spaces, and an architectural ethos that seamlessly embeds preventive health into daily routines rather than discarding it as a mere afterthought.

THE ECONOMY OF ATTENTION...

Amid the glare of screens and the constant ping of notifications, modern life has become a battleground for attention (Gazzaley & Rosen, 2016). Offices bristle with open-plan layouts, offering zero refuge from incessant noise, while public spaces morph into commercial arenas, saturating visitors with advertisements and digital messages (Carr, 2011). Although technology often bears the blame, architectural design plays a substantial yet underexamined role in stoking these fires of overstimulation and burnout (Christakis & Zimmerman, 2007).

An emerging field of "attention architecture" seeks to understand how built environments either aid or erode the ability to focus. Research suggests that overstimulation arises when design elements flood human senses without allowing for organic breaks or restful zones. In workplaces, open-plan office spaces have grown ubiquitous, championed for promoting collaboration and efficient real-estate usage (Pouwels, 2020). Yet these grand halls of desks provide little or no acoustic privacy, minimal visual barriers, and a perpetual hum of chatter and foot traffic (Pouwels, 2020). Workers struggle to sustain concentration, their attention hijacked by each ring of a colleague's phone or the swirl of movement in peripheral vision. Over time, such relentless exposure can spawn chronic stress, diminished job satisfaction, and an uptick in mental fatigue (Huang et al., 2022).

In parallel, public spaces once conceived for communal gathering—such as shopping centers, airports, and transit hubs—are increasingly designed to serve commercial imperatives rather than human comfort. Such design choices reflect a broader societal shift toward perpetual stimulation as a driver of consumer engagement, aligning with marketing strategies that aim to seize every spare second of the occupant's attention (Smith & Jones, 2019). People navigating these hyper-commercial spaces experience an elevated baseline of tension, a result of constant sensory bombardment that undercuts reflective thought.

The detrimental effects of this overstimulation are not purely anecdotal. Multiple studies link high-sensory environments to adrenal fatigue, anxiety, and even compromised decision-making (Huang et al., 2022). Furthermore, the body's stress-response system is evolutionarily tuned for acute threats, not for persistent, low-level sensory distractions, color, and motion. When there is no structural reprieve—no configurable zone, no "quiet island" of respite—humans remain in a subtle but persistent state of alertness. Over weeks and months, cortisol levels rise, attention spans shorten, and individuals become prone to burnout (Huang et al., 2022). Similarly, attempts to carve out quiet zones in commercial spaces often falter because the entire blueprint is oriented toward maximizing foot traffic or exposure to advertising content. Mall operators, for example, frequently weigh the commercial opportunity cost of installing seating clusters for relaxation against the revenue gained from an additional kiosk.

Architects and occupational psychologists have begun calling for "focus-centric design," advocating for adjustable partitions, flexible furniture arrangements, and strategic use of acoustical materials. These measures, though beneficial, often clash with corporate interests or zoning regulations that hold fast to monolithic floor plans. Even the concept of a dynamic office that transforms throughout the day—quiet in the morning, collaborative at noon, then hush again by late afternoon—runs afoul of mechanical infrastructures that are neither agile nor easy to retrofit.

Beyond hardware, the software of occupant behavior matters as well: social norms shape whether employees feel comfortable wearing noise-cancelling headphones or leaving an open floor for a more private zone. If the design lacks explicit cues that "quiet is welcome here," cultural and managerial pressures may discourage workers from seeking solitude. The end result is a built environment that implicitly demands continuous engagement—no matter how cognitively exhausting that might be.

At the heart of this predicament lies an "economy of attention," a term referencing how modern societies treat human focus as a scarce commodity exploited for profit (Simon, 1971). The architecture of overstimulation dovetails neatly with an economic system that prizes moment-to-moment engagement, either from employees in an office or consumers in a mall or individuals doomscrolling and online shopping at home. But the toll on mental health is steep. As the WHO notes, burnout is increasingly recognized as a legitimate occupational phenomenon (WHO, 2019). This crisis underscores the link between unyielding physical settings combined with ceaseless digital engagement, and the resultant depletion of cognitive and emotional resources. Reimagining architecture as a sanctuary of mindful design stands as a significant challenge—yet also a beacon of hope. Early adopters of "low-stimulus" office concepts report improved employee satisfaction, reduced absenteeism, and enhanced innovation.

In sum, overstimulation and digital burnout are not mere byproducts of technology but direct consequences of architectural choices that fail to adapt to human physical, psychological, and social rhythms. By hardwiring spaces in ways that saturate the senses, we invite persistent stress, short-circuiting the mind's ability to focus or find respite. As attention becomes the new currency, the unadaptable building emerges as a Trojan horse of endless stimuli, leaving those inside no place to hide—until, eventually, the meltdown arrives. Resisting this tide calls for a radical shift in design ethos: we must craft rooms and halls that gently cradle the human mind rather than wringing it dry.

SCENARIO 7: OVERSTIMULATING OFFICE SPACES

Paulette steps onto the polished concrete floor of Auracore Tower, an office complex revered in the media for its sleek, "collaborative" layout. But from the moment she enters, it feels like stepping into a neon fishbowl. Rows of open desks stretch infinitely under harsh LED lighting. Glass partitions reflect meeting pods crammed with high-gloss posters, while overhead, a digital scoreboard blinks with employee metrics.

Her friend Damon greets her at a standing desk. "Welcome to the hive," he mutters. "We have no walls, only data." He taps the scoreboard that ranks employees by daily tasks completed—a corporate experiment, or so management claims. Paulette sees colleagues glancing at the scoreboard every few minutes, pulled from their tasks by the lure of competitive metrics. No one can hide in this environment.

They settle at Damon's station. Behind them, an artificial waterfall gurgles. "Isn't that supposed to be relaxing?" Paulette asks. But as soon as a meeting down the hall ends, a new wave of chatter merges with the waterfall's droning rush, forming a cacophony of white noise.

Office drones scuttle from one side to another, transfixed by digital signage scrolling corporate announcements. Overhead, translucent screens occasionally flash brand slogans: STAY EFFICIENT. STRIVE HARDER. Paulette tries to type on her laptop, but her attention flickers each time someone passes by. The row of desks is too narrow to avoid glimpses of every phone call, every micro-gesture. "It's chaos," she whispers, scanning for any hint of an escape.

Architecture's silent vow: I've locked you into a single plane of collaboration.

In a corner, a meager "privacy booth" stands—just big enough for one occupant. A queue forms for its claustrophobic interior, yet people rarely emerge calmer. Damon gestures toward it. "If you want silence, you get a phone-booth prison. That's the trade."

At midday, the floor's chatter becomes a deafening hum. Each desk is a battlefield for the occupant's concentration. Paulette rubs her temples. "I can't think," she confesses. Surges of text messages, Slack pings, and overhead announcements invade every moment. The building itself offers no respite: no modular or reconfigurable furniture, no modular partitions or reconfigurable walls. Auracore Tower is a single, monolithic design.

She tries the cafeteria, hoping to find a quieter corner. Instead, she's confronted by overhead digital menuboards with pulsing animations. Even there, large screens tout performance metrics. Workers slump in chairs, their eyes flitting from phone to screen, an aura of tension saturating the air.

When the day ends, Paulette's nerves feel frayed. As she exits, she passes a corporate plaque touting "Open Space for Open Minds." The irony stings. In this environment, minds feel locked in open spaces, battered by overstimulation they cannot flee.

SCENARIO 8: THE HYPER-COMMERCIALIZED PUBLIC SPACE

Later in the week, Paulette meets her friend Lucia at Technomall Nexus, a colossal shopping center connected to a major transport hub. The entrance glitters with jumbo LED arches, each pulsing with shifting advertisements— new phone deals, instant-credit promotions, and a swirl of multi-colored logos. Lucia stands near a fountain that glows electric purple from submerged lights.

"Did you see the signs for the relaxation lounge?" Lucia asks, ironically. The lounge turns out to be a narrow row of lounge chairs adjacent to a two-story digital billboard blaring the next brand campaign. There's no partition, no acoustic barrier. So-called "relaxation" merges with the endless commercial feed.

They roam the complexity of walkways lined with shops. Glossy floors reflect the glare of overhead lighting. Neon signage competes for every second of eye contact: BEST OFFERS, LIMITED TIME SALE, HURRY NOW. Lucia points out an overhead speaker piping promotional jingles every few minutes. "No wonder I feel on edge," she sighs. "They want me to stay alert so I keep shopping."

> Architecture's silent vow: I have no quiet corners. Every surface is an advertising canvas.

Crowds push forward. Families juggle strollers while loudspeakers ring out train departures for the adjoining train station. The building's design channels foot traffic through a central corridor lined with pop-up shops. Lucia tries searching for a restroom, but the signage is buried among brand posters. By the time they find it, a wave of travelers emerges from the terminal, funneling into the mall with suitcases in tow.

"Look at that," Paulette says, pointing at a mezzanine that could have hosted a quiet seating area. Instead, it's a gallery of digital screens, each rotating marketing messages. People stand, transfixed or disoriented, scanning the screens for gate updates or product deals. The swirl of color, motion, and sound forms a vortex with no calm center.

They attempt to exit through a side corridor, hoping for respite, only to find it cordoned off by a kiosk for the latest tech gadget. The kiosk blasts pop music on loop. Lucia's shoulders tense. "There's no real way out but to pass more shops," she groans, caught in the cyclical design that forces them back into the main retail zone.

At last, they reach the exit leading to the train station. The corridor is lined with vending machines that beep incessantly, advertising snacks on small screens and digital banners advertising destinations to visit. Even here, digital signage wraps columns, urging them to come back soon. ENJOY YOUR NEXT EXPERIENCE AT TECHNOMALL NEXUS!

Lucia glances at Paulette, eyes weary. "I just wanted to pass through, but it feels like we spent all our energy warding off the mall's assault on our senses."

Stepping into the cooler evening air, they pause on a sidewalk bathed in the glow from the building's exterior façade. Without the bombardment of neon, they both exhale, as though emerging from a dream of synthetic chaos. "It's staggering," Paulette reflects, "how architecture can be used to keep us captive, not just physically but psychologically."

Lucia nods. "We might survive it, but we pay with our peace of mind. Next time, let's meet in a park."

They share a brief laugh, but the tension lingers. In the reflection of Technomall's mirrored panels, the city's ephemeral lights scatter across the night sky—unapologetic in their brilliance, wedded to a design that harnesses every second of attention. A fleeting thought crosses Paulette's mind: What if we had spaces built for restoration, not stimulation? What if spaces were designed to maximize our agency over how we experience them? What if they were designed with our best interests in mind? For now, that remains a dream overshadowed by the relentless hum of commerce and the architecture that sealed it in place.

TECHNOLOGICAL UNEMPLOYMENT AND THE SKILLS GAP...

In the past decade, rapid shifts in technology—from automation to AI-driven analytics—have upended traditional workplace models (Brynjolfsson & McAfee, 2014). Once, office buildings operated as static monoliths, comfortably hosting nine-to-five employees whose schedules and roles were predictable (Evans & McCoy, 1998). However, the advent of remote work, hybrid arrangements, and continuous education demands a more flexible approach to architecture—one that can pivot swiftly in response to labor-market and societal needs (Davenport & Beck, 2001). Those structures that remain rigid become ghosts of an earlier era, half-filled or fully vacant, their hushed corridors and deserted meeting rooms emblematic of wasted potential and economic inefficiency (Frey & Osborne, 2017).

Many observers argue that the meteoric rise in remote work was an inevitable outcome of digital transformation, merely accelerated by global disruptions such as pandemics or supply-chain crises (Barrero et al., 2021). Yet the conversation rarely extends to the physical buildings themselves—those once bustling office towers that now stand with entire floors dark, HVAC systems underutilized, and reams of commercial space collecting dust (Evans & McCoy, 1998). The unadaptability layer emerges when buildings, constructed for rigid occupant capacities and single purposes, cannot retrofit for communal use, startup incubators, or skill-building hubs (Sundstrom et al., 1994).

Underpinning this shift is also a ballooning skills gap: as industries adopt advanced AI, robotics, and data-driven processes, workers require ongoing retraining (Autor, 2015). Traditional educational facilities remain locked into an older mold—lecture halls for single-instructor classrooms, labs with outdated equipment, and schedules ill-suited for fast-paced skilling (Darling-Hammond, 2010). Such spaces cannot easily integrate new teaching technologies or collaborative methods like project-based learning and remote group work. Lacking adjustable furniture, robust Wi-Fi infrastructures, and modular partitions, these environments hinder the creativity and agility that modern learners need (Evans, 2003).

This dual challenge—abandoned offices and outdated educational facilities—reflects a deeper structural friction: architecture conceived in an industrial age collides with a digital, fluid workforce (Florida, 2017). Sociologists emphasize that when physical spaces block skill acquisition or hamper new work models, social inequalities multiply (OECD, 2018c). Urban cores brimming with vacant high-rises fail to harness the potential of hybrid communities seeking coworking nodes, digital labs, or cross-generational skill-sharing spaces (Glaeser, 2011). Meanwhile, underfunded schools remain locked in the architecture of the past, unable to outfit themselves for tomorrow's jobs (Darling-Hammond, 2010).

Observing this phenomenon leads to a critical question: How might architecture evolve to align with the dynamic needs of a changing workforce? Some developers propose "conversion corridors," retrofitting office towers into multi-use complexes that host coworking floors, adult-learning centers, and skill-up pods (Evans & McCoy, 1998). Others advocate for modular design from the outset—buildings where walls can shift, communal areas can expand or contract, and entire floors can reconfigure from office space to classroom to VR lab overnight (Davenport & Beck, 2001).

Economic forces also impede swift change. Property owners often fear diminished property values if a "corporate tower" is rebranded as a hybrid coworking-education facility (Frey & Osborne, 2017). Yet in the face of rising labor flexibility and the unstoppable march of AI-driven disruptions, many experts argue that clinging to single-purpose spaces equates to planned obsolescence (Susskind & Susskind, 2022). The mismatch between building design and workforce requirements not only wastes resources but can also stifle local economies, leaving entire districts hollowed out (Cohen, 2020). Interestingly, some urban planners argue that abandoned office towers could become the solution to outmoded school facilities: through adaptive reuse, these structures might morph into dynamic skill centers, bridging the gap between education and on-the-job training (Evans & McCoy, 1998). However, these transformations remain sporadic, often championed by niche developers or specialized nonprofits, as mainstream institutions resist complexities in financing, building codes, and structural retrofits.

At the social level, observers note that architectural inertia contributes to labor displacement (Acemoglu & Restrepo, 2018). As automation eliminates certain roles, employees require spaces to retrain or incubate entrepreneurial ventures. If they find none—because their offices remain locked in a nine-to-five flux and local schools have no agile, high-tech labs—joblessness can escalate, fueling social unrest. In essence, the built environment's unresponsiveness amplifies the shockwaves of technological upheaval.

A wave of proposals emerges from the confluence of these tensions. Some researchers advocate for "flow-based zoning," allowing a building's classification to shift with occupant needs. Others push for "education-credit building codes," awarding tax incentives if property owners designate floors for training facilities, coworking spaces, or vocational labs. Real-world pilots show promise: a half-abandoned office tower in Berlin was reconfigured into a continuous-learning center with flexible micro-classrooms and coworking pods that adapt to occupant usage via sensor-driven partitions. The building's foot traffic soared, local unemployment dipped, and property values stabilized—a testament that architectural agility can yield tangible socioeconomic benefits.

Still, challenges loom. The future workforce demands spaces that can accommodate daily metamorphoses—multi-generational learners, remote teams converging for a week, AI-driven startups needing ephemeral labs, or novices exploring new trades in pop-up training zones. Achieving such fluidity necessitates not merely a rethinking of materials and partitions but also a cultural shift among architects, city officials, and real-estate developers (Evans & McCoy, 1998).

In conclusion, the tension between old, single-purpose buildings and the new demands of a technologically disrupted workforce underscores a key truth: architecture is not merely a backdrop but an active player in shaping economic and social outcomes (Evans, 2003). Without spatial and architectural evolution, half-empty offices and rigid schools will continue to stand like relics—static monuments to an era that prized uniformity over resilience.

SCENARIO 9: ABANDONED SINGLE-USE OFFICE SPACES IN THE ERA OF REMOTE WORK

Paulette steps off the commuter rail and into MarbleRay Tower, once the pride of the city's corporate skyline. Its lobby flaunts polished floors, an echoing hush where footsteps resonate. She's meeting her friend Nadia, who's scouted a vacant floor for a tech incubator. But they've heard the place is locked into archaic design codes.

As she enters, the building's AI announces in a clipped synthetic voice:
   "Architecture (MarbleRay):
   'Welcome, occupant. Please scan your corporate ID.'"
   Paulette, who works freelance now, tries to press a "Visitor Access" icon. The system denies her, repeating: Corporate ID not detected. Undeterred, Nadia meets her in the lobby, swipes an old badge, and leads Paulette to the 24th floor.
   All around them, evidence of a vanished era: hallways lined with uniform cubicles, silent but for the faint buzz of overhead lights. "Remote work changed everything," Nadia murmurs. "Most staff left. Now these floors stand empty."

MarbleRay Tower (in a discreet whisper from an overhead speaker):
   "This space is zoned for professional occupancy, Monday to Friday, 9 AM to 5 PM. Deviations require property manager approval."
   Nadia sighs. "We want to host a code bootcamp here. Teach AI fundamentals, web dev, maybe rent coworking desks after hours." She flicks a light switch, but half the ceiling fluorescents remain dark. "The building's wiring is on a timer locked to old schedules. Outside those hours, entire sections go offline."
   Paulette glances through a row of dust-coated windows: an urban panorama with silhouettes of newer, nimbler buildings. "Why not adapt? The location's perfect."

MarbleRay Tower (soft mechanical tone):
   "My design architecture prohibits structural modifications. Unapproved usage is restricted by municipal code 418-B."
   They step into the main office zone—rows of identical desks, each with a dust film on neglected monitors. The place feels deserted, haunted by an intangible memory of corporate bustle. Nadia tries to set up a modular partition for a makeshift training corner but finds anchor points in the floor missing. The building was never intended for flexible reconfiguration.
   They hear footsteps behind them. It's Jerome, a property manager. "Look, I'd love to see this place used," he admits, "but the owners want prime corporate tenants. The city code says this is 'Class A Office Space.' You can't just host an educational program."
   Nadia clenches her fists. "So it stays empty while job seekers scramble for training?" She gestures at the wide, vacant floor. "It's a travesty."
   MarbleRay Tower remains inert, the old AI offering no solution.
   A week later, they try to hold a small pilot event anyway. Computers hum, participants gather. The overhead lights flicker—some parts of the floor remain

locked out. Mid-session, the AC system cuts off because the building's schedule perceives no "corporate presence" after 5 PM. Participants endure stifling heat, forging ahead despite the stench of stale air.

Paulette looks around, sweat on her brow. "We could do so much here," she murmurs. "If only the architecture allowed real-time adaptation."

At 6 PM sharp, an automated shutter clangs down in the corridor, sealing off an entire wing. Students scramble to gather materials. The system's unyielding nature forcibly ends the session.

MarbleRay Tower (final, echoing broadcast):
   "Office hours concluded. Have a pleasant evening."
   Paulette and Nadia exchange exasperated glances. They step into the deserted hallway, the building's illusions of grandeur overshadowed by a stifling design that belongs to a past that refuses to let go.

SCENARIO 10: EDUCATIONAL FACILITIES OUTPACED BY TECHNOLOGICAL ADVANCES

A month later, Paulette finds herself at AshWood Academy, an aging institution known for its majestic brick façade and long corridors. Her friend Roland, a teacher transitioning into AI-based curricula, asked her to visit. "We're stuck in the 20th century," he warned. "We need help upgrading."

Upon arrival, Paulette marvels at the antique architecture: tall, narrow windows, wooden desks bolted to the floor. The hallways smell of chalk and disinfectant. From hidden speakers, a muffled announcement drifts, instructing students to line up for midday assembly. She meets Roland in a barren corridor leading to the "IT lab," a cramped room of outdated desktop PCs.

AshWood Academy (via a static-laced overhead intercom):
   "Please maintain single-file movement in corridors."
   Roland rakes his fingers through his hair. "I proposed an AR/VR lab for advanced skill training, but the building can't handle the wiring or ventilation needed. Also, the desks in our classrooms are anchored in rows—no group work or reconfigurable setups."
   They peer into a typical classroom: battered chairs in neat lines, a dusty chalkboard at the front. The single overhead projector squeaks as it swivels. "We want to run robotics clubs," Roland laments, "but there's no floor space. The building codes say we can't remove these built-in seat risers without structural rewiring. The cost is insane."
   Paulette spots a teacher battling with ancient Wi-Fi routers near the ceiling. "We tried hooking up advanced telepresence gear," the teacher says, exasperated. "The building's walls block signals, and we can't open them up because they're load-bearing. The entire structure is like a fortress for an era that's long gone."

AshWood Academy (slightly crackling intercom):
"Take your seats promptly. Classes commence in five minutes."
Roland gestures toward the hallway leading to a defunct auditorium. "We wanted to transform that space into a collaborative lab for programming and design. But the seats are fixed in an amphitheater format, plus the stage is a relic with limited power outlets. We can't run half the equipment we need."
They pass a cluster of bored students outside the library—a library stocked mostly with outdated textbooks. The students stare at phone screens. One quietly remarks, "We have to watch online tutorials because the labs don't have the software we need." Paulette wonders how many bright minds remain under-stimulated here, lacking an environment conducive to modern skill-building.
Roland leads her to a locked door labeled "Experimental Annex," rumored to hold leftover lab equipment. Inside, dusty 3D printers lie idle. Overhead lighting flickers. "We tried hooking these up for a Maker Club. The building's circuit breaker kept tripping. Now the printers gather dust, victims of an electrical grid designed for overhead projectors, not advanced fabrication."

AshWood Academy (an archaic building computer pipes in, monotone):
"Maintenance request needed to activate Annex power. Estimated wait time: six months."
Paulette shakes her head in disbelief. "It's 6 months to rewire a single room? Meanwhile, students graduate unprepared for the technologies shaping tomorrow's industries."
They exit into an outdoor courtyard, overshadowed by tall brick walls. The open sky offers the only breath of modernity. Roland sighs, glancing at the century-old structure. "This place is beloved for its heritage. But nostalgia won't equip kids for AI-driven workplaces. We need flexible reconfigurable furniture, AI-driven labs, advanced bandwidth. Instead, we're locked into a relic."
As they walk away, a handful of students gather under a battered oak tree, trying to connect to the building's sporadic Wi-Fi. Paulette senses their frustration: the environment itself stands in the way of learning. She imagines them as future job-seekers, left behind because the building literally cannot adapt.
That evening, Roland confides, "We petitioned to convert the old library into a co-learning space, but local preservation laws fought us. They claim changing the interior structure is a violation of historical integrity." Paulette sighs, reflecting on how architecture, once revered for tradition, now stifles the emergence of new skill sets.

AshWood Academy (final announcement for the day):
"Classes concluded. Remain within designated zones."
They exit through heavy wooden doors, the echo of footsteps following them down the corridor. Outside, the modern city hums with 5G or probably 6G signals and digital screens. Within, the Academy stands as a silent witness to a previous century's constraints. In that moment, Paulette feels the weight of how built environments can chain entire generations to outdated learning, dimming the promise of new opportunities.

SOCIAL FRAGMENTATION AND POLARIZATION...

In contemporary cities, social inequality manifests not only in economic statistics but also in the stark physical landscapes that delineate who belongs and who does not (Florida, 2017). Far from neutral backdrops, developments increasingly adopt rigid layouts—whether for efficiency, security, or high-end appeal—that effectively segregate communities along income or cultural lines. Studies of gentrification demonstrate how working-class neighborhoods are transformed by real estate speculation and redevelopment (Lees et al., 2013). Displacement follows, as long-time inhabitants cannot afford rising rents and are forced to relocate to areas lacking comparable services. Meanwhile, these luxury structures resist any reconfiguration toward mixed-income housing or genuinely shared amenities, entrenching a landscape of social fragmentation.

Public spaces, once democratic hubs for cross-cultural and intergenerational exchange, have similarly been sterilized. As urban sociologist Jan Gehl emphasizes, the best public realms allow adaptation—pop-up markets, street performances, impromptu gatherings—but many modern designs prioritize clear circulation for shoppers or office workers over human interaction (Gehl, 2011). The result is thoroughfare-like environments where people pass through rather than pause to connect. These layouts not only canalize pedestrian flows for security or profit but also limit opportunities for spontaneous mixing across socioeconomic divides.

The psychological toll of exclusionary architectures is profound. Residents in areas devoid of welcoming public amenities or who feel unwelcome in luxury developments report heightened alienation and resentment. Even well-intentioned cultural institutions can falter. A new library or community center, if sited behind restrictive gates, set apart from transit routes, or arranged in inflexible layouts, may be physically accessible yet psychologically off-limits to many. Conversely, centers designed with modular rooms, movable furniture, and transparent façades invite diverse groups to claim space for festivals, workshops, or social gatherings—practices shown to build mutual understanding and reduce tensions.

Nevertheless, there are exemplars of inclusive design. In some cities, citizen coalitions have successfully lobbied for mixed-income mandates in new developments or the preservation of shared plazas in redevelopment plans. Adaptive reuse projects have converted vacant towers into co-housing communities with sliding scales of rent, communal kitchens, and rooftop farms. Such initiatives demonstrate that when architecture is imagined as mutable—capable of evolving with community needs—it becomes a catalyst for solidarity rather than division.

SCENARIO 11: GENTRIFICATION AND EXCLUSION IN URBAN DEVELOPMENT

Quarstone Quarter—an upscale district that was once an industrial neighborhood. Now, glimmering condo towers flank a manicured canal, overshadowing the modest row houses that survived the transformation.

Paulette arrives by tram, stepping into a plaza where polished granite pavers reflect the midday sun. Her friend Marcus, a longtime resident, greets her near a decorative fountain that spurts water in perfect arcs.

"Welcome to what they're calling the 'New Renaissance,'" he says wryly. "And they do mean new, because nearly everything old got bulldozed."

They wander toward the condominium entrances, each adorned with wrought-iron gates. Above, sleek balconies boast lush rooftop gardens—private to owners only. A discreet security camera swivels, scanning visitors.

Architecture (Quarstone Tower) in a smooth, AI-driven voice:
"Access restricted to residents or authorized guests. Please present your key card."

Paulette attempts to peer inside the lobby, glimpsing marble floors and a lounge plush with velvet seating. "You used to walk here freely?" she asks Marcus.

He nods. "Back when this was a textile warehouse. We held community events, open markets. Now you can't even step in without an invitation. The design's made exclusion official."

They stroll along the canal, passing restaurants with chic terraces. The menus show steep prices, a subtle barrier for lower-income locals. A newly installed footbridge is blocked by an electronic gate—"private walkway," reads a sign.

Quarstone Tower drones again from hidden speakers:
"Please respect property boundaries. Violators subject to ejection."

Marcus sighs. "I grew up here. Now I feel like an intruder." He points to a block of older row houses overshadowed by the tower. Residents have hung protest banners from windows: STOP FORCED RELOCATION. Their voices echo, overshadowed by the hum of renovation machinery.

Paulette notices that the entire promenade is lined with uniform benches bolted to the ground at exact intervals, all facing the condo façade. There's no flexible seating for gatherings, nowhere for kids to play spontaneously. "Everything's choreographed to look pristine but feel ... sterile," she observes.

Marcus: *"Gentrification turned this place into a curated stage. Locals got forced out, or they sold under pressure. The architecture locks us out physically, and psychologically."*

They pause at a small playground—only it's behind another wrought-iron fence labeled Residents' Amenities. A child presses her face against the bars, longing to join. "It's not a public park," Marcus whispers. "She can't go in."

"Paulette shakes her head, hearing an undercurrent of electronic hum from the building's security system. She imagines it murmuring: Keep out. This is not for you. As they depart, the tower's polished façade gleams with the promise of elite living, an opulent fortress proclaiming, Only some can pass.

SCENARIO 12: ABSENCE OF INCLUSIVE COMMUNITY SPACES

One week later, Paulette meets Naomi and Darius at Newton Square, a city-run park known for its curious mix of wide paths and poorly designed communal areas. On approach, they find a symmetrical grid of concrete walkways with minimal shade. A few static benches line the perimeter, each angled away from one another, offering no chance for group conversation.

Architecture (Newton Square), via an automated kiosk at the entrance:
"Welcome. Please note that large gatherings require prior approval. No reconfiguration of park furniture is allowed."

Naomi gestures at the kiosk. "We wanted to host a neighborhood potluck, but they said we can't move benches or set up communal tables."

Darius points to a fenced-off section designated "Sports Area." It's a single basketball court with tall, locked gates. "You need to reserve it through an online system," he says. "Also, no changes allowed—no net for volleyball, no open space for group yoga."

They wander through the main plaza. In the center stands a sculpture encircled by an immovable bench in a perfect circle. It looks aesthetic but isn't comfortable. Elderly folks hover at the edges, unsure where to rest. Young kids race around the sculpture, lacking a dedicated playground. Darius tries to pick a shaded spot, only to realize the layout offers no refuge from the blazing sun.

Architecture (Newton Square), kiosk repeating a monotone advisory:
"This park is for passive recreation. Please maintain order and respect local noise guidelines."

Paulette raises an eyebrow. "Passive recreation? So basically, no spontaneous events." She notices two teenagers on skateboards eyeing a ramp-like structure. But the sign warns: No skateboarding allowed. They leave, disappointed.

Naomi recounts a story: "Last month, a group tried to hold a multicultural festival here. They wanted a stage, some tables for food stalls. But the city said the design doesn't support large gatherings. The benches can't be moved, and the kiosk's system flagged it as 'excessive use of public space.' So they gave up."

A single mother with a toddler stands near a locked restroom with a Closed for Maintenance sign. She looks around helplessly, not finding a family-friendly corner. "An inclusive park would've had child-sized seating, a playground, a reading nook," Naomi laments.

Darius: "This place lacks soul. It's just ... wide pathways, rigid benches, and those weird surveillance cameras on posts. Feels more like a sculpture garden than a community space."

          Paulette sits on a bench, frowning as the bright overhead sun stings her eyes. "Maybe the architects thought minimalism was elegant. But they forgot about real people's needs. There's no shading, no flexible seating, no cultural expression. Everybody's forced to either comply or leave.

"They spot a group of seniors attempting to gather in a corner. Two want to push benches together, but the benches are bolted down, their arrangement *purely ornamental. They give up, scattering into the hot sunlight. The park's emptiness betrays how the design disinvites communal life.*

Architecture (Newton Square):
"We hope you enjoy your stay. Please refrain from unauthorized modifications to park furnishings."
Naomi shakes her head. "This is no conversation zone, no cross-cultural mingling. It's a pass-through space, encouraging nobody to linger."
Paulette stands, the kiosk's voice echoing in the distance. She envisions a lively fair with tents, music, and families relaxing on movable chairs. But the kiosk's rigid policies and the park's fixed layout make that impossible. As they leave, an electronic sign flashes new updates: Upcoming City Event: Private Function. The park remains reserved for an exclusive event, ironically closed to the general public.
In that moment, Paulette feels the weight of architecture's silent decree: only certain forms of social engagement are permitted, strictly regulated, limiting spontaneity and diverse encounters. If community cannot gather, divisions inevitably harden elsewhere. The park's design, though cloaked in neat orderliness, quietly stifles the intangible bonds that might knit people together.

THE FUTURE OF WORK...

Contemporary shifts in the labor market—fueled by digital platforms, real-time analytics, and global connectivity—have propelled the gig economy to the forefront (De Stefano, 2016). Once dismissed as a marginal side-hustle, gig work now constitutes a central pillar of modern employment (Kalleberg & Dunn, 2016). Alongside this evolution, the demands on physical spaces are transforming: workers no longer confine themselves to nine-to-five routines in fixed offices; instead, they drift—seeking adaptable corners in cafés, co-working hubs, or even pop-up "micro offices" carved out of vacant storefronts (Manyika et al., 2016).

On the home front, the situation is equally dire for many gig workers lacking proper remote-work facilities. Traditional apartments, conceived in an era that drew a hard line between "live" and "work," seldom accommodate the complexities of home-based jobs (Bloom & van Reenen, 2015). Simultaneously, modern equipment—large monitors, specialized lighting, advanced teleconferencing systems—demands infrastructure upgrades that older buildings cannot support without costly retrofits (McKinsey Global Institute, 2019a). Consequently, remote gig workers endure suboptimal conditions: hunched over laptops in cluttered corners, competing for space amid domestic clutter, or battling incessant background noise (Kalleberg & Dunn, 2016). Moreover, the absence of suitable home-office amenities exacerbates mental-health challenges. Gig workers face social isolation, wage instability, and the relentless pressure of 24/7 connectivity (Standing, 2011). When forced to work in environments ill-suited for concentration—where poor ergonomics or family distractions persist—stress accumulates. Research shows that environments devoid of natural light, comfortable seating, or minimal noise lead to reduced productivity and heightened burnout risk (Gajendran & Harrison, 2007), underscoring that architecture's unyielding framework carries significant psychological costs (Rosen, 2016).

This dissonance between the fluid, ephemeral nature of gig work and the monolithic stance of traditional architecture speaks to a broader oversight: the role of the built environment in shaping economic opportunities (Susskind & Susskind, 2022).

Technology, paradoxically, both accelerates and partially mitigates this crisis. Gig platforms proliferate, driving an exponential increase in workers seeking agile spaces (Manyika et al., 2016), while advanced AR/VR solutions promise "virtual co-working" that facilitates remote collaboration in digitally rendered environments. However, even these immersive tools cannot substitute for the need for a quiet, physically comfortable space for extended tasks; humans remain tethered—at least for now—to the constraints of desks, chairs, and immovable walls.

A philosophical re-examination of "home" and "workplace" is also imperative (Sennett, 2007). As the boundaries between work and home blur in the gig economy, architecture might reconceptualize homes as hybrid spaces—part public, part private—equipped with flexible partitions that allow living areas to transform into professional zones without major renovation. In conclusion, the gig economy signifies a seismic shift in the nature of work—one that demands equally radical innovations in architecture (Brynjolfsson & McAfee, 2014).

SCENARIO 13: LACK OF CO-WORKING OPTIONS FOR GIG WORKERS

Solaris District is a city block once renowned for its corporate towers that have partially emptied as remote work surged. Paulette arrives to meet her friend Elena, a freelance graphics designer juggling short-term contracts.

Paulette: *"So this is the rumored co-working floor?"*
Elena: *"Supposedly. But rumor has it they lease it only to large startups for big money. Not exactly 'open' co-working."*
             They step inside, greeted by a polished lobby with an art installation of neon rods overhead. A sign directs them to "Venture Suite – By Appointment Only." Elena sighs. "I emailed them. The monthly membership costs more than my rent."

ApexOne (the building's automated system chimes):
    "Welcome. Please note that unauthorized entry to Venture Suite is prohibited. Day passes are unavailable."
    They ride an elevator to a deserted floor. Rows of empty offices line the corridor, each locked. Through glass walls, they glimpse immaculate desks, pristine potted plants, and a lounge area with coffee machines.

Paulette: *"All this, sitting idle, while gig workers crowd cafés without proper seating."*
             Elena lightly taps the glass. No response.

ApexOne (over speakers):
    "Access restricted. This facility is designated for corporate affiliates only."
    They move toward an "interim lounge" and find a solitary table cluttered with promotional brochures, sparse seating, dim lighting, and a notice reading "Private events only." Elena checks her watch, anxious to deliver a video pitch in half an hour. She sets up her laptop, cursing the echo off the hard floors and the faint hum of an AC vent.

Elena: *"The acoustics are terrible. Clients will hear every sound."*
            A security guard appears, frowning. "This lounge is not public co-working. Please vacate if you're not authorized."
Elena: *"I just need fifteen minutes."*
Guard: *"Sorry, management policy. No exceptions."*
            Paulette and Elena are ushered back to the lobby. Outside, they observe a cluster of gig workers huddled over laptops near a coffee kiosk, awkwardly spread out.
Paulette: *"So many vacant floors, yet so little actual space for us."*
Elena: *"We're stuck in limbo—not large enough to rent a suite, not upscale enough for a membership."*
            As they leave, the building's reflective façade gleams in the afternoon sun. Elena muses, "Maybe we could try the library or that café on Third Street." Yet both know those spots lack stable Wi-Fi or quiet corners for confidential calls. ApexOne looms behind them, an unyielding fortress that spurns the flexible worker's reality.

SCENARIO 14: INSUFFICIENT AMENITIES FOR HOME-BASED WORK

Two weeks later, Paulette visits her friend Malik, a data analyst juggling multiple gig contracts. He rents a cramped studio in Riverview Towers, a dated residential building notorious for its antiquated design.

Malik: *"Welcome to my 'office,'" he jokes, gesturing to a small table wedged between his sofa and kitchenette. He's rigged a second monitor on a stool.*
            Paulette notices tangled cables, a flickering overhead bulb, and laundry draped on a foldable rack near his "workspace."
Paulette: *"So you have no dedicated area? Don't you struggle with background noise during conference calls?"*
Malik: *"All the time. The walls are paper-thin—if my neighbor plays music, my clients hear it. Plus, there's no space for acoustic panels."*
            They venture into the hallway in search of a common lounge or rec room. A sign points to "Community Hall," but they find it permanently locked with a notice stating "Reserved for building events

only. Must schedule through management." Malik shakes his head. "I tried once; they said it's not for private use. Even if I wanted to host a co-working session, the furniture is fixed in an auditorium style."

Riverview Towers (a robotic female voice from the hall speaker):
"Residents must not alter communal areas. Unauthorized modifications are prohibited."
Malik leads Paulette to the rooftop, which is partially accessible. He once dreamed of a makeshift open-air workspace with fresh air and quiet, but instead, they find a fenced area cluttered with outdated deck chairs and a defunct barbe-cue station.

Malik: *"I asked if we could clean it up, maybe install shared tables, but man-agement said it would breach structural codes. The building's design never envisioned a true rooftop lounge."*
              They survey the view: a scenic river below, overshadowed by dis-tant steel high-rises.
Paulette: *"You can't get Wi-Fi up here either?"*
Malik: *"No—the signal's patchy. I even tried a booster, but the thick walls block it."*
              Returning inside, they pass neighbors similarly frustrated. One mutters about wanting to open an in-home yoga studio but is thwarted by the fixed floor layout that can't handle extra foot traffic or minor expansions. Another complains about archaic wiring that trips if more than two electronics run simultaneously. Malik's living space epito-mizes larger problems: no acoustic insulation, no possibility for parti-tioning, no flexible storage for office.
              Exasperated, Malik slumps at his tiny desk as he prepares for a virtual meeting.
Malik: *"I lose hours every day just coping with my environment. This build-ing anchors me to outdated assumptions—like everyone leaves at 8 AM for an office, so who needs a proper workspace at home?"*
              Paulette picks up a flyer from the building manager's desk touting "Modern living for the modern professional!" She scoffs at the irony.
Paulette: *"They forgot that modern professionals might actually work from home—and need more than just a corner."*
              With a resigned shrug, Malik dons his headset. The looming pile of laundry threatens to tumble into view on his webcam—a stark reminder of how his space fails to adapt to the new rhythms of gig work. Through the open window, the distant hum of traffic seeps into his call.

Riverview Towers (distant overhead announcement):
"Reminder: Noise policy is in effect after 9 PM."
Paulette watches the meeting commence, aware that architecture remains the silent antagonist—confining Malik in a cramped battleground between domestic life and professional demands. She wonders how many others in this building share his plight, longing for flexible spaces that simply do not exist.

HUMANS, ARCHITECTURE, AND THE DIGITAL DILEMMA...

Cities stretch across the horizon like intricate circuit boards etched on the planet's skin, shimmering under an overcast sky pulsing with the subtle hum of data streams. Structures tower where outdated ideas of walls and windows have given way to integrated networks crafted more for control than communion. Interlaced within these environments is a restlessness—a clear sign that the era of single-purpose architecture has collided with the multifaceted demands of human existence. In that collision, modern inhabitants find themselves forced into confined niches, compelled to twist their lives to conform to spaces that no longer reflect how they work, learn, or dream.

This symptom is visible everywhere: a bustling business district filled with pristine towers, their marble lobbies so polished that footfalls echo like ghosts, now dimmed after traditional hours (Brynjolfsson & McAfee, 2014). Unadaptable structures, once celebrated for their efficiency, now lie half-lit, processing only the memory of a workforce that has scattered into remote, hybrid, and itinerant modes (Manyika et al., 2016). Meanwhile, individuals seeking quiet havens for their fleeting labor wander these corridors in vain. The building's skeleton remains too rigid for digital nomads who might only stay half a morning and who crave flexible pods to spark transient creativity. The tension extends beyond professional realms. In residential blocks, families attempt to transform living rooms into partial offices, creating makeshift nooks behind folding screens that barely mute the neighbor's murmurs or the persistent beep of a child's game console (Gajendran & Harrison, 2007). Over time, this friction intensifies into quiet frustration as inhabitants are forced to adapt to the dwelling's inflexible geometry rather than having a design that adapts to their evolving routines.

At the same time, entire districts remain unprepared for environmental volatility: storms, heat waves, and floods intensify year after year (McKinsey Global Institute, 2019c). Buildings fitted with single-purpose mechanical systems and poorly conceived thermal envelopes become oppressive ovens when air conditioning falters under extreme loads. Rooflines cannot readily support solar expansions or green installations. Along storm-prone coasts, water infiltrates these rigid structures, leading to catastrophic failures that expose the arrogance of an industry that once prized formulaic repetition over adaptive resilience (Brynjolfsson & McAfee, 2014).

Within public realms, overstimulation sculpts another form of suffering. Commercial zones engineered with blinding LED panels, looping audio announcements, and fixed walkways funnel visitors along predetermined routes, leaving no alcove for reflection or modular refuge for those seeking quiet (Rosen, 2016). This continuous bombardment not only erodes the psyche but also breeds isolation, as individuals become less inclined to interact in spaces designed more for consumer seduction than for communal bonding (Spinuzzi, 2012).

Debate around social fragmentation intensifies as urban enclaves, once symbols of inclusivity, now primarily serve the affluent; their guarded entrances and elevated skywalks create physical barriers from working-class neighborhoods (Kalleberg & Dunn, 2016). Instead of nurturing a tapestry of diverse interactions, these enclaves are molded by single-use logic: private pools, exclusive lobbies, membership-only rooftop terraces. Residents within these walls seldom encounter the cultural mosaic beyond, and the architecture silently executes a sorting mechanism that reinforces economic fractures (Florida, 2017). The cumulative outcome is not always overt catastrophe—but a quiet, persistent suffering: bodies pressed into unyielding corners, communal ties slowly fraying as no forum for spontaneous exchange emerges, and a mounting anxiety as climate extremes strike without resilient built responses (McKinsey Global Institute, 2019c). In aggregate, these micro-sufferings become the signature afflictions of the modern built environment.

Yet, glimmers of possibility flicker on the horizon. In some quarters, architects, civic planners, and grassroots organizations champion multi-functionality as the antidote to rigidity (Brynjolfsson & McAfee, 2014). Advocates of these new designs also highlight adaptability in materiality: composite walls that can be reconfigured, modular acoustic panels that detach and reattach with ease, and lighting systems that mimic natural circadian cycles (Rosen, 2016). Entire floors might be reassigned to new uses with minimal expense, bridging the gap between fleeting trends in remote work or collaborative learning.

At the urban planning scale, adaptive design can address the digital dilemma—where technology's relentless input meets architecture's static stance. Some conceptual frameworks propose digital-free sanctuaries, akin to "architectural quiet zones" that ban invasive ads and interactive signage (Rosen, 2016). This challenge calls not only for technical innovation but for a fundamental shift in values—from short-term profit and aesthetic monumentality to long-term resilience and occupant-centered design.

As we turn the page from this part of pervasive pain to a new horizon, we introduce Part IV: *Rays of Hope/Heroes of the Near Future*—a beacon emerging from the twilight of despair. Here, the dialogue shifts from the relentless weight of suffering to the luminous promise of transformation. In this forthcoming section, we shall explore how innovative systems, adaptive architectures, and human ingenuity converge to forge new pathways for resilience and renewal. The journey ahead invites us to witness the reimagining of our built environments, the coalescence of human and machine, and the spirited defiance of those who dare to dream of a future where hope is not a distant ideal, but a vibrant, tangible reality. As we embark on this next part, let us embrace the vision of a world where suffering is met not with despair, but with the transformative light of human creativity and adaptive collaboration.

As Paulette walks home one evening, she reflects on her encounters with friends and colleagues over the past several weeks and she notices something interesting. No matter where she is, the communication technology all around her is ubiquitous, overstimulating, incessant, everchanging, and prevents socializing and well-being. Conversely, the built spaces themselves are not stimulating enough, are not configurable/reconfigurable at all, and also prevent socializing and well-being. It appears that communication technology must be quieted whereas the architecture must be enlivened—and both need to better support socializing and well-being. The two respective, complementary sets of technologies must meet in the middle and perhaps integrate to best support people at work, at home, and when socializing. She wonders how this might be done.

Part IV Rays of Hope/ Heroes of the Near Future

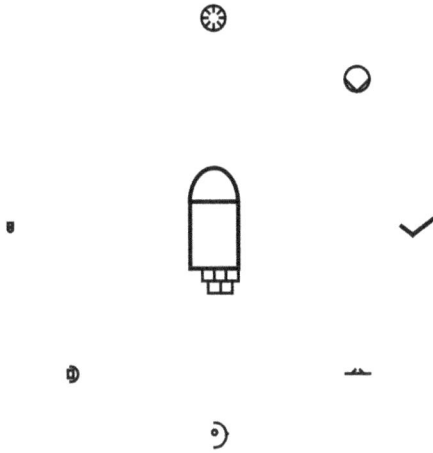

PORTAL TO THE FUTURE...

Human societies are entering a liminal season in which the familiar logics of industrial modernity falter while planetary risks surge and technological capacities accelerate. Within this moment of transition, there is a need to substantially alter how the built environment is designed, constructed, and inhabited (Brynjolfsson & McAfee, 2014). In a time of transition, single-purpose artifacts—whether an office tower, a school, or a zoning code—become obsolete quickly; that is, they are obsolete well before the ends of their useful lives. What is to be done with a building designed to last 100 years that is obsolete in 25 years? During such periods of transition, value lies in cultivating architectures that, like biological ecologies, privilege adaptation, support resiliency, and mitigate stressors. In order to achieve such benefits in the era of Industry 6.0 technology, fundamental advancements are required for how we design, construct, and inhabit the built environment.

The demand for adaptability, resiliency, and stress mitigation is visible first in the economic sphere. Global platform labor, remote collaboration, and on-demand production have splintered the 19th-century schedule of factory time into networks of asynchronous micro-tasks (De Stefano, 2016). Yet most workplaces remain configured as rigid grids of cubicles and fixed meeting rooms, spatial scripts that tacitly enforce managerial surveillance, linear workflows, and nine-to-five occupancy.

DOI: 10.1201/9781003441953-5

The discrepancy between how space is configured versus how it needs to be configured to support activities causes and/or exacerbates stress. *Empirical meta-analyses of telecommuting show that productivity and psychological well-being hinge less on digital tools than on whether workers possess agency to modulate acoustic, thermal, and social boundaries throughout the day* (Gajendran & Harrison, 2007). Options already exist to grant such agency, but adoption is throttled by financing models that reward lowest-cost standardization over long-term resilience (McKinsey Global Institute, 2019a).

Spatial inertia is equally consequential in public health. The COVID-19 pandemic revealed that hospitals optimized for high patient throughput under stable conditions can become powerful transmission nodes when pathogen profiles shift (Dietz et al., 2020). Rapid retrofits—machine-learning–guided triage, pop-up negative-pressure micro-wards, and outdoor recovery gardens—were practicable only where earlier masterplans had set aside programmable floor plates and redundant service shafts for contingencies (Chong et al., 2021; Li & Nishikawa, 2022).

Philosophically, such flexibility resonates with Deleuze's ontology of multiplicities, wherein entities unfold as flows rather than fixed essences (Deleuze, 1992). Bernard Tschumi's program for "cross-programming" anticipates this stance by treating each building as a stage for conflicting, overlapping, and evolving activities (Tschumi, 1996). When designers script contradiction into initial layouts—oversized service cores, plug-and-play structural grids, demountable façades—they seed capacity for what resilience scholars term adaptive migration: the orderly relocation of functions without abandonment of assets. The same logic animates biologists' view that species-rich ecosystems withstand perturbation better than monocultures because latent niches can be rapidly occupied after disturbance.

Psychological research further confirms that human flourishing thrives in environments that offer both refuge and prospect—zones of sensory calm adjacent to arenas of stimulation (Rosen, 2016). Trauma-Informed Design (TID) extends this gradient, specifying daylight gradients, tactile materials, and volumetric variation that lower cortisol levels among occupants who have experienced stress or displacement. Meta-reviews show consistent correlations between biophilic cues and improved cognitive performance, immune response, and social cohesion (Ulrich et al., 2008). Accordingly, any truly future-ready system must embed TID strategies and biophilia as adjustable infrastructure: living walls clipped into façade cassettes, circadian lighting tuned by occupant biodata, and acoustic baffles grown from mycelium that can be composted and replaced as program shifts.

In a way, this next evolutionary step in Evidence-Based Design (EBD) serves as a precursor to when architecture extends beyond Earth. The amount of initial and ongoing validation and verification required for designing, constructing, commissioning, and operating intelligent, interactive environments on Earth is similar to what will be required to sustain human-habitable environments away from Earth. By way of example, consider the amount of sensing, perception, and processing power required for the International Space Station, which contains only a few thousand square feet of living and working space (i.e., the size of a large, upper middleclass U.S. suburban home). Per NASA's fact sheet (NASA, 2024) on the ISS, its operation requires:

1. approximately 8 miles of wiring,
2. 350,000 sensors, "...ensuring station and crew health and safety,"
3. More than 50 computer control systems,
4. More than 3 million lines of code, and
5. 100 data networks processing more than 400,000 signals that track vital metrics.

Imagine a near future wherein the built environment on Earth approaches this level of complexity in order to provide intelligent, adaptive architecture employing the use of Industry 6.0 technology. To do so requires a fundamental and significant expansion of the knowledge bases brought to bear in designing and operating the built environment. The design of residences, schools, and commercial spaces will exceed the complexity and rigor of today's state-of-the-art industrial and medical facilities. And yet designing for this level of complexity is here already—but so far just limited to the aerospace, defense, and nuclear industries, with some near equivalents in laboratory design, semiconductor facility design, biotech/pharma production facility design, data center design, autonomous storage and retrieval systems design (i.e., robotic warehouses), hospital design, and other advanced manufacturing facilities design. As shown in the coming pages, Ultra-Large-Scale Systems (ULSS) are a foundational project type for realizing intelligent, adaptive systems.

Long-duration habitation on the Moon or Mars, or long-term space travel, will couple extreme environmental constraints—radiation, vacuum, isolation—with equally extreme demands for adaptability, resilience, and stress mitigation. Habitability studies demonstrate that sensory monotony and social claustrophobia, rather than mechanical failure, are primary predictors of mission burnout (Spinuzzi, 2012). Modular inflatable shells, robotic regolith printers, and closed-loop life-support systems must therefore interface with immersive simulation rooms, flexible privacy pods, and garden biomes to sustain mental health. Lessons from such extraterrestrial laboratories can reciprocally retrofit terrestrial cities, where rising sea levels and heat waves redraw habitability gradients with analogous suddenness.

Ethics enters this conversation because built systems are never neutral; they distribute risks and affordances unevenly. Ethical architecture must therefore pair technical flexibility with governance mechanisms that guarantee access: community land trusts, inclusionary zoning for multifunctional shells, and open-source platforms for participatory retrofitting (Gensler, 2020a; Gensler, 2020b). Such frameworks transform inhabitants from passive tenants into co-developers, aligning maintenance decisions with lived expertise and reducing the rebound effects of top-down masterplans.

Critically, the shift from architecture as object to architecture as ULSS does not imply endless fluidity divorced from identity. Empirical studies of workplace redesigns caution that perpetual reconfiguration without narrative anchors can breed disorientation and territorial conflict (Manyika et al., 2016). The challenge is dialectical: provide stable scaffolds—structural grids, service spines, cultural rituals—while allowing secondary layers to modulate at temporalities matching social and operational changes.

If such synthesis appears ambitious, pilot projects already demonstrate feasibility. Singapore's public-housing blocks have integrated void-deck retrofits—community kitchens by day, flood shelters during monsoon events—supported by adaptable MEP trunks that anticipate future climate thresholds.

In conclusion, standing before the portal to the future means acknowledging that yesterday's monuments of permanence have become tomorrow's stranded liabilities. By integrating cyber-physical feedback loops, TID, biophilia, and ethically distributive governance, architecture can evolve from a static backdrop into an active mediator of human and planetary well-being. The remainder of this part summarizes some key constructs and methods for developing adaptable, resilient, and stress-mitigating environments as ULSS.

BUILDING BLOCKS...

RAY I: AUTOMATED SYSTEMS

Internet of Things

The IoT is defined by the International Telecommunication Union (ITU) as a global infrastructure for the information society, enabling advanced services by interconnecting physical and virtual things based on existing and evolving interoperable information and communication technologies (ITU, 2012). IoT is a component of CPS. In architecture, IoT has revolutionized building interactions with occupants and the environment. Smart buildings now integrate IoT sensors to monitor parameters such as air quality, temperature, occupancy, and structural health. The Edge in Amsterdam exemplifies this integration, utilizing approximately 28,000 sensors to monitor lighting, temperature, humidity, and occupancy, thereby optimizing energy efficiency and occupant comfort (Randall, 2015). IoT architectures in buildings typically consist of layered systems: embedded sensors collect real-time data, communication modules transmit this data via networks; and cloud or edge computing platforms analyze and store the data (Miorandi et al., 2012). This structure supports applications like intelligent traffic systems and predictive maintenance, where systems proactively report malfunctions. However, the proliferation of interconnected devices raises concerns about security and interoperability. Standards developed by organizations such as IEEE and National Institute of Standards and Technology (NIST) are crucial for addressing these challenges (Sicari et al., 2015).

IoT is somewhat of a generalist term. In reality, there are a number of formally defined emerging project types that embody the dynamics of complex, interactive architectural systems. As shown in Figure 4.1, this chapter will summarize most of these. The common elements of these emerging system types are the following:

1. A component of a larger complex/interactive system of systems while being composed of systems of systems;
2. Real-time hardware/software interactions among and between internal and external systems to function successfully; and
3. Real-time human-machine-software interactions are essential to meeting user goals and expectations (Manganelli, 2013).

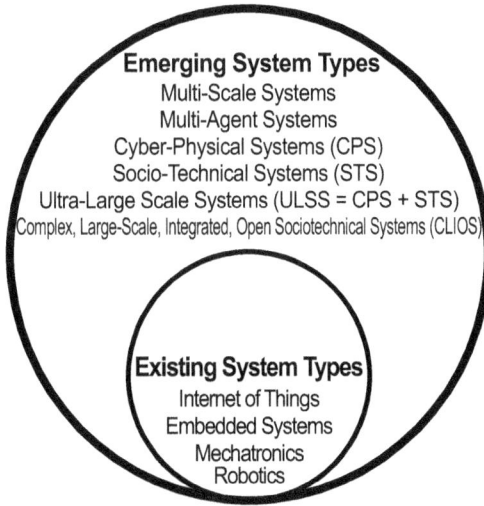

FIGURE 4.1 There are several emerging project types that entail greater scale, complexity, integratedness, and real-time interactivity that will come to define how we use and design environments in the near future (Manganelli, 2013).

Cyber-Physical Systems

CPS represent the integration of computation, networking, and physical processes, enabling embedded devices to monitor and control real-world phenomena through continuous feedback loops (Lee et al., 2015; see Figure 4.2). CPS are "globally virtual and locally physical" (Rajkumar et al., 2010; Xie, 2006). In architectural practice, CPS have evolved from isolated building-automation components into distributed, cloud-linked ecologies that imbue façades, structures, and urban infrastructure with real-time responsiveness. A leading, empirically validated exemplar is Barcelona's Media-TIC building, whose pneumatic shading façade—composed of over 4,500 air-filled cushions—modulates daylight admission and solar gain by responding to exterior irradiance and temperature measurements. Field studies report energy savings of up to 20% compared with conventional fixed-shade systems, demonstrating the measurable impact of CPS-driven envelopes on energy performance and occupant comfort (Abitare, 2010).

The Al Bahar Towers, designed by Aedas Architects, feature a dynamic façade composed of umbrella-like panels that open and close in response to the sun's movement. This responsive system reduces solar gain and glare, enhancing occupant comfort and decreasing reliance on air-conditioning systems. The integration of sensors and actuators enables the façade to adapt in real time to environmental conditions, exemplifying the application of CPS in kinetic architecture. Another example is the Mood Swing project, an affective kinetic building façade system that employs sensor-based automation to respond to external environmental conditions. This system utilizes sensors to detect changes in the environment and actuators to adjust the façade accordingly, demonstrating the potential of CPS to create responsive and adaptive building envelopes.

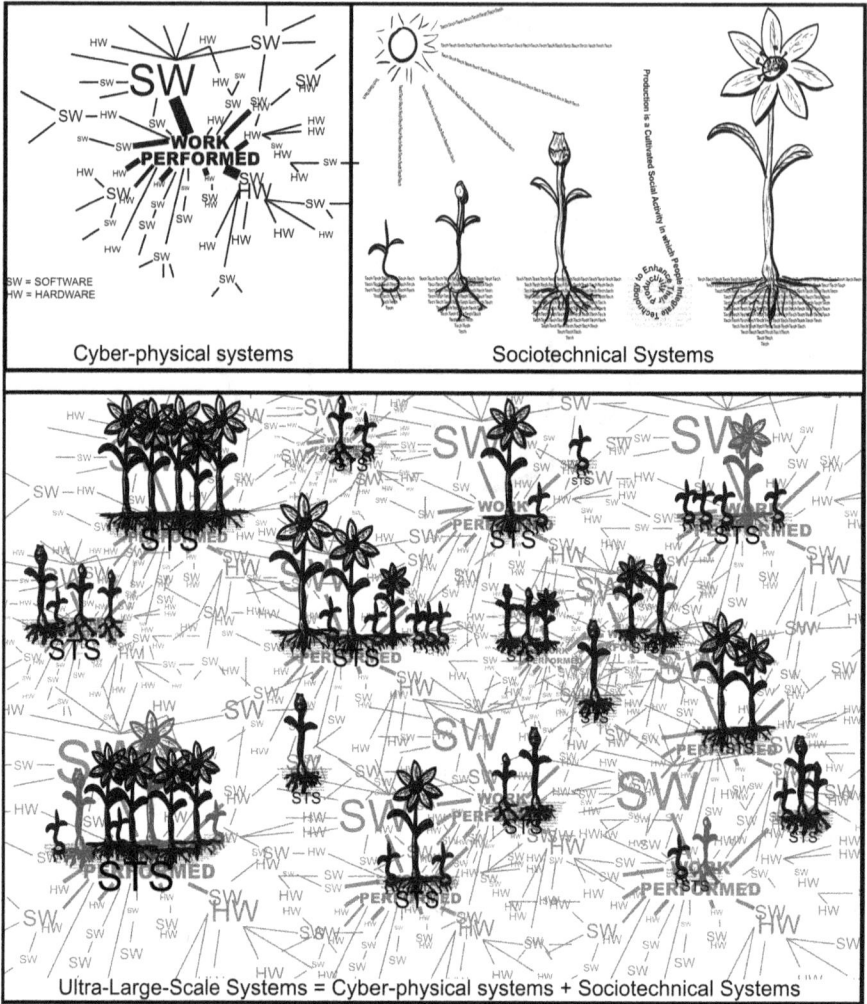

FIGURE 4.2 CPS integrate layered software and hardware systems aiding user tasks (Manganelli, 2025). STS involve human-centered processes that must be cultivated through use, not predesigned. ULSS arise as STS within pervasive CPS environments—systems of systems that evolve through continual cultivation (Manganelli, 2015).

Ecodistricts, in which utilities are shared among a neighborhood in order to realize greater efficiencies due to economies of scale, also represent the level of systems integration requisite for CPS. Lastly, NIST produced a CPS Framework guide document (Griffor et al., 2017) and a community-led assessment of how to design and operate what it called Internet-of-Things-Enabled Smart Cities (IES-City) that addressed how to design CPS (Bhatt et al., 2018). It codified its findings in the IES-City Framework. The IES-City Framework defined its principal goal as lowering barriers to interoperability between systems, as doing so lays the groundwork for achieving large-scale, integrated, open systems.

As CPS scale from individual buildings to metropolitan platforms, their architectures layer edge devices for high-frequency sensing, local controllers for millisecond actuation, fog nodes for medium-latency analytics, and cloud services for long-term pattern mining and predictive scheduling (Lee et al., 2015). This multiscale topology underpins applications ranging from adaptive street lighting to resilience-driven water management. Yet formidable challenges persist: millisecond-class responses demand deterministic networking and precise clock synchronization, while actuator-level cyber-intrusion could weaponize a kinetic façade or smart bridge—necessitating real-time encryption, zero-trust authentication, and robust intrusion-detection protocols (Rajkumar et al., 2010).

Despite these challenges, field evaluations clearly demonstrate the benefits of CPS: empirically observed energy savings of 20–50%, accelerated post-disaster service recovery, and significant improvements in occupant satisfaction and productivity (Abitare, 2010). As edge AI, resilient network topologies, and comprehensive cybersecurity measures mature, CPS has the potential to transform buildings from passive structures into active, adaptive agents—advancing urban resilience, optimizing resource utilization, and enhancing human well-being.

Sociotechnical Systems

STS were first developed in the late 1940s for analyzing industrial work processes that require integrations of skilled human teams with task-specific technologies. STS theory maintains that technology components serve as work process armatures that structure how work is done. Characteristics of STS include:

1. STS will have to be adapted in field, iteratively, as they integrate with the social structure and organizational dynamics;
2. the social component cannot be specified and must be incrementally evolved; and,
3. the symbiosis between technology and social dynamics optimizes the STS.

Achieving a well-functioning STS entails the following characteristics:

1. The *work system* is the basic organizational unit;
2. The *work group* is all people involved in a *work system*;
3. The work is *internally regulated*;
4. The STS, including its people, exhibit *redundancy of functions*, so that no one person or piece of technology is a point of systemic failure;
5. The team cultivates *discretionary* (flexible) work roles for individuals;
6. The human team and the machines are *complementary*; and,
7. The STS is *variety-increasing*, making a more flexible, resilient system.

Strategies for designing STS include the following:

1. a *culture of participation* with an *ecology of roles*;
2. users adapt procedures and tools over time;

3. intentionally *under-designing* intended work culture and supporting systems so that the *under-designed* elements evolve/optimize practice, proce-
   dure, and supporting systems within the team while in use;
4. identifying infrastructural elements that must be specified without participatory design versus those elements which may be adaptable for the user
   while in use;
5. *semi-structured modeling*, in which the system design coevolves with the
   participation of stakeholders; and,
6. *walk-through-oriented facilitation*—that is, *under-designed* interventions iteratively cultivated into fully functioning components of the STS
   (Herrmann, 2009; Fischer & Herrmann, 2011).

In summary, the design of STS requires iterative co-design of the elements because
not all immediate and future uses can be understood during the design phase (see
Figure 4.2).

Ultra-Large-Scale Systems

The aerospace and defense industries in particular have pioneered the concepts ger-
mane to how to design large, integrated, systems of systems. In particular, in the early
2000s, the U.S. Army commissioned the Software Engineering Institute (SEI) to
assess how to design and operate an integrated human/organization/machine system
of systems that takes "...billions of lines of code to run..." (Northrop et al., 2006).
SEI produced a definition of a new project type called ULSS. *ULSS are composed of
a combination of networks of CPS and STS* (see Figure 4.2).

Importantly, the ULSS report acknowledged that it was not possible to fully
design, engineer, and specify a ULSS in advance of its use. Rather, like all STS, core
technological components of the ULSS must be engineered, placed in their context
of use, given to the end-users, and then the end-users must operate the system, and
the remaining design and optimization of the ULSS must occur as an iterative feedback loop between end-users, organization, and technical systems, with the system in
operation. *That is, ULSS cannot be fully specified in advance. Rather, they must be
cultivated into existence.*

There are similar precedents with the design and operation of existing advanced
manufacturing facilities. For the design and operation of biotech/pharma, semiconductor, and other advanced manufacturing facilities, the systems start up and commissioning, qualification, and validation (CQV) processes can take as long or longer
than the design and construction of the facility—for example, it is common that a
modern biotech/pharma production facility takes 1–3 years to design, 2–4 years to
build, and 3–5 years to startup and commission. As we move toward a world composed of ULSS, and then we develop systems for habitation in space, these types of
CQV processes will become essential for the startup and optimization of all human
built environments with Industry 6.0 technology.

Complex, Large-Scale, Integrated, Open, Sociotechnical Systems

CLIOS are complex, large-scale, integrated, open, sociotechnical systems. CLIOS
address nested complexity, especially when CPS are distributed across wide geographic regions, and therefore subject to compliance with multiple policy and

regulatory frameworks, some of which are identical, some of which are complementary, and some of which conflict. CLIOS can be thought of as ULSS that exist within a complex, multilateral, and distributed policy and regulatory environment, which adds an extra layer of administrative and regulatory burden to the design and compliance of the ULSS. In summary, CLIOS model systems for which "…the degree and nature of the relationships is imperfectly known, with varying directionality, magnitude and time-scales of interactions" (Sussman, 2007).

Human-AI-Robot Teaming

Human-AI-Robot Teaming (HART) (Holder et al., 2021), Human-AI-Teaming (HAIT or HAT) (Pflanzer et al., 2022), Human-Machine-Teaming (HMT) (Damacharla et al., 2018), and Human-Systems-Integration (HSI) (Booher, 2003) are terms used to describe scenarios in which task completion is dependent on real-time coordinated activity by humans and machines, including but not limited to AI and robots. A key concept is that sensing, perception, cognition, and action are distributed across team members, and all team members must work together to achieve the goal. Team members may be biological agents or non-biological, software or machine agents, i.e., biotic agents or abiotic agents. HMT and HSI are traditional terms for these types of systems of systems analyses and design, whereas HART, HAIT, and HAT are newer terms. This text uses the HART term for convenience. The important point is that the sensing, perception, cognition, and action produced by the HART system are distributed across all agents. Given this, it should be clear that this means of engaging the world is an instance of combining CPS+STS. In other words, HART defines a class of subsystems within a ULSS.

SAE Autonomy Scale and a Framework for Multi-Agent Interactions

The Society of Automotive Engineers developed a scale for levels of automation (Society of Automotive Engineers (SAE), 2021) that is increasingly being adapted for use in other design domains. The levels of automation are:

Level 0 (Manual):
All functions remain under direct human control, with the environment serving as a responsive tool for user-driven operations.
Level 1 (Assistance):
Limited automation augments human effort by optimizing discrete tasks— such as temperature regulation or lighting adjustment—while retaining manual oversight.
Level 2 (Partial):
Repetitive functions, including cleaning and routine maintenance, are automated, though human supervision remains essential for overall system integrity.
Level 3 (Conditional):
The system operates autonomously under predetermined conditions, for instance, managing energy conservation during peak usage periods, while permitting human intervention if necessary.
Level 4 (High Automation):
In most scenarios, the system proactively orchestrates environmental parameters with minimal need for human input, reserving manual overrides solely for exceptional anomalies.

Level 5 (Full Autonomy):
A fully self-governing system that anticipates and executes all operational requirements, seamlessly integrating into the user's lived experience.

Given that SAE Levels of Automation Scale are being adapted for other industries, it is likely that they will be adapted for design, construction, and operation of the built environment. From this perspective, it is likely that design of intelligent, adaptive environments will entail identifying different levels of automation for different systems and subsystems. For instance, while a concrete stoop may have level 0 or 1 automation, conversely, a main entry door may warrant level 4 automation. It is reasonable to conceive of designing a building wherein each component in the BIM has a parameter to assign its level of automation. With respect to vehicle automation, there are four types of agent communication and interaction for Autonomous Driving Systems (ADS): V, V2V, V2I, and V2X (Yusaf et al., 2024). Vehicle (V)— an AI-enhanced agent or agents that is/are part of a single vehicle for perception, cognition, decision-making, and/or action. Vehicle to Vehicle (V2V)—AI-enhanced agents that participate in a local swarm or hive, wherein the onboard agents in a vehicle benefit from the collective perception, cognition, decision-making, and/or action of the other vehicle-based agents around them and coordinate behaviors with them. Vehicle to Infrastructure (V2I)—an AI-enhanced agent or agents that is/are part of a vehicle that communicates with road-infrastructure agents to enhance perception, cognition, decision-making, and action. Vehicle to Everything (V2X)—an AI-enhanced agent or agents that is/are part of a vehicle that communicates with all other autonomous agents around them, as well as remote "central" monitoring/on-the-loop agents, for the most comprehensive perception, cognition, decision-making, and action. This same break down is applicable for any human-AI-robot-teaming system of systems —- and as a starting point for the likely nature of interactions between non-human agents in intelligent, adaptive environments. So we generalize to this set of constructs: Agent (A), Agent-to-Agent (A2A), Agent to Infrastructure (A2I), and Agent to Everything (A2X). This agent network exists within an Operational Design Domain (ODD), with each agent engaging with the ODD according to its own Operating Envelope Specification (OES) (Griffor et al., 2021). The performance and teamwork of the human and non-human agents are coordinated per the Human Readiness Level (HRL) scale (See, 2021). It is reasonable to conceive of designing a building, wherein each component in the BIM has parameters and constraints for A, A2A, A2I, A2X, ODD, OES, and HRL.

Industry 4.0, Industry 5.0, Industry 6.0

Industry 4.0, as defined by the European Commission, refers to the integration of CPS, IoT, and cloud computing to create intelligent, connected systems that enable real-time process optimization and product customization (Lasi et al., 2014). In the built environment, Industry 4.0 principles extend to influence design, construction, and building management. Vertical integration ensures seamless data flow from sensor-equipped materials to building management systems, while horizontal integration connects supply chains, facilitating collaboration among architects, contractors, and facility managers (Kagermann et al., 2013).

Industry 5.0 augments Industry 4.0 technologies by establishing human-centered design and operation as a core requirement. Focusing on human-centeredness improves systems performance overall by improving the human components of the system to reduce error rate, improve reaction time, reduce cognitive and physical workload for users, improve overall performance metrics, and make for a more resilient process.

Industry 6.0 augments Industry 5.0 technologies and human-system integrations by increasing the pervasiveness of such technologies, continuing to improve the human-centeredness aspects of task execution, and by incorporating additional emerging technologies like augmented reality, so that systems are self-optimizing in close to real time (Das & Pan, 2022).

As these technologies evolve, the influence of Industry 6.0 is expected to permeate urban environments, reshaping residential areas and municipal infrastructure to create responsive, resilient cities. *Collectively, ULSS and Industry 6.0 systems represent foundational elements of intelligent, adaptive environments.* By embedding digital intelligence into physical structures, these systems enable buildings to respond dynamically to environmental changes, occupant needs, and unforeseen events. This convergence enhances efficiency and sustainability, enriching the human experience and paving the way for resilient urban ecosystems capable of thriving amid the complexities of the 21st century. But we must recognize that the way to design and develop such systems requires a paradigm shift. It is no longer the case that humans can completely design a system in the abstract, build it, and then run it. Rather, through their iterative use and refinement, *humans must cultivate ULSS and Industry 6.0 technologies into being.*

RAY II: HUMAN SYSTEMS (SCIENCE OF WELL-BEING)

Psychofortology

Psychofortology, derived from the Greek "psyche" (mind) and Latin "fortis" (strong), is a subset of positive psychology, focusing on psychological strengths and resilience. This field emphasizes innate human capabilities that enable individuals to thrive amid adversity, examining constructs such as optimism, hope, adaptive coping strategies, and robust social support networks as critical determinants of well-being (Coetzee & Cilliers, 2001). The six principles of psychofortology (Coetzee & Cilliers, 2001) are:

1. sense of coherence,
2. locus of control,
3. self-efficacy,
4. hardiness,
5. potency, and
6. learned resourcefulness.

Just a quick perusal of this list makes plainly clear that the current use of much telecommunications, social media, internet, and security technology violates all of the principles of psychofortology. It is no wonder that these technologies impart so much psychological trauma.

Hedonic Psychology

Hedonic psychology is the scientific study of pleasure, happiness, and the pursuit of positive affect, focusing on how momentary positive emotions and overall life satisfaction are influenced by internal dispositions and external conditions (Kahneman et al., 1999). Central to this field is the concept of the hedonic treadmill, which posits that individuals tend to return to a stable baseline of happiness after significant positive or negative events, highlighting the role of adaptation in sustained well-being (Brickman & Campbell, 1971).

In architectural contexts, hedonic psychology provides a framework for designing spaces that actively enhance pleasure and well-being. Buildings like the Bullitt Center in Seattle are designed not only for energy efficiency but also to maximize occupant satisfaction through features such as expansive windows, natural ventilation, natural materials, and green spaces, promoting a sense of calm and contentment (Hanford, 2014).

Public spaces that integrate diverse sensory experiences—ranging from water features and natural landscaping to thoughtfully arranged public art—offer residents opportunities to experience delight and engage emotionally with their surroundings (Gehl, 2010). Research indicates that environments enriched with aesthetic and sensory stimuli can stimulate creativity, increase social cohesion, and enhance cognitive performance (Vartanian et al., 2013).

Multiple Resource Theory (MRT)

Wickens (1984) developed the 4-D multiple resource model and MRT, which describes how information can be distributed across perceptual channels to optimize perception, cognition, and response. The MRT 4-D model has the following dimensions:

- stages of processing: perception, cognition, and response;
- codes of processing: a spectrum of processing from spatial cognition to verbal cognition;
- modalities of perception: information distributed across auditory and visual channels;
- visual channels: separates use of focal and peripheral vision.

To the basic model is added a model of resource allocation (Navon & Gopher, 1979) that demonstrates by calculation that information presented on different channels can be processed more efficiently than the same amount of information all presented through one channel. With respect to development of technology and the built environment, MRT helps predict how to distribute information presented to users across channels of perception such that the most information can be perceived and used with the least resource usage.

Trauma-Informed Design

TID is an approach that incorporates an understanding of trauma's lasting impact on individuals and communities into the creation of physical spaces and workflows, emphasizing safety, trust, empowerment, collaboration, and choice to minimize environmental triggers and foster healing (SAMHSA, 2014). The five principles of TID are:

1. Views of nature;
2. Varied lighting strategies;
3. Residential finishes;
4. Minimal clutter; and
5. Autonomy of control.

In the built environment, TID principles are applied across various settings. Hospitals and mental health facilities often feature open, airy layouts with ample natural light and designated quiet zones to reduce anxiety and facilitate recovery (Stokols et al., 2013). Social service agencies and shelters implement features such as secure entry points, clear sightlines, and sound-insulated rooms to create a sense of safety and control (Berg & Upchurch, 2007).

Research demonstrates that environments informed by trauma-sensitive principles can positively affect mental health outcomes. Educational settings incorporating TID have shown reductions in behavioral issues and enhancements in academic performance by providing students with safe, supportive spaces (Stokols et al., 2013). Workplaces adopting TID principles often see improvements in employee well-being and reductions in stress-related absenteeism. TID is also being applied to mitigate the stress and harm of social media and other technology-induced traumas (Scott, et al., 2023). As cities address the long-term effects of both natural and man-made disasters, TID offers a vital framework for creating resilient communities where physical spaces contribute directly to psychological healing and recovery.

Collectively, psychofortology, hedonic psychology, MRT, and TID provide an integrated, wellness-based and empowerment-based framework for reimagining human systems interactions within the built environment. By grounding architectural practices in these evidence-based principles, designers can create spaces that both meet functional requirements and also actively foster resilience, elevate social and personal well-being, and support recovery from trauma. Such environments become instrumental in shaping a healthier, more vibrant society—where the built environment serves as both a sanctuary and a catalyst for human flourishing.

RAY III: SYSTEMS VALIDATION

Requirements Engineering

RE is defined by the IEEE as the systematic process of eliciting, documenting, and managing the needs and expectations of stakeholders to ensure that a system or product meets its intended purpose (IEEE, 2017). In essence, RE is about understanding what stakeholders require, both in terms of functionality—what a system must do—and non-functional aspects such as performance, security, usability, and reliability. Importantly, a key part of RE is *validating* the requirements. That is, just because a stakeholder says something does not mean that it is true. In fact, typically, the majority of what stakeholders say is noisy and cannot be validated. Thus, the validation process brings tremendous value to system design.

Imagine designing an intelligent, adaptive hospital patient room. The stakeholders give the requirements engineer over 1,400 need statements (i.e., potential requirements). But during the validation process, only about 300 of the need statements can be validated as true and necessary. These 300 requirements become the basis for design. The design challenge is much more manageable and cost-effective—and likely to yield much better results—because time and energy are not spent designing features and functions that are not really needed. In fact, using a rigorous RE process, it is common to reduce the number of requirements by 70–95% compared to just assuming that all of the stakeholder statements are valid.

Typically, in the context of the built environment, RE is applied to software systems but not to the design of the building. Sadly, proper RE methods are mostly unknown in architectural practice. However, RE is complementary to EBD and enhances traditional standard architectural best practices for space programming methods. Therefore, there is ample opportunity to improve architectural space programming by incorporating proper RE practices.

The process begins with an *exploratory* elicitation phase, during which engineers and architects engage directly with stakeholders—ranging from city planners, building occupants, and community members to private investors—using interviews, surveys, workshops, and direct observation. At the end of the exploratory phase, there is a set of potential requirements, but it is not yet clear which ones are valid. Existing architectural best practices for space programming typically stop here, assuming the validity of what stakeholders tell the design team, rather than going through the process of validating the stakeholder's stated concerns.

Following the elicitation phase, the next step in RE is a series of *confirmatory* requirements analyses used to *disambiguate (i.e., make orthogonal and measurable), validate, categorize, and prioritize* the requirements (Sommerville & Sawyer, 1997). Techniques used include but are not limited to: use case modeling, syntactic and semantic disambiguation, qualitative and quantitative analyses, open and closed card sorts, follow-up surveys, Delphi Analysis, Kano Analysis, scenario analysis, and the development of user stories and personas. These techniques help to resolve uncertainty and conflicting thoughts among stakeholder demands and ensure the appropriateness of the design. The output is a set of clear, actionable, validated requirements that serve as the blueprint for the project.

Documentation is a critical component of RE. All requirements are formally recorded in structured documents such as Software Requirements Specifications or Business Requirement Documents. In the field of architecture, similar documentation might include detailed design briefs, such as the Owner's Project Requirements and Basis of Design documents as defined by ASHRAE Guideline 0 (ASHRAE, 2019). These documents act as a reference for developers, architects, testers, and all stakeholders throughout the project lifecycle, ensuring that the evolving design remains aligned with the original vision.

Validation refers to whether or not we are even designing the right thing. Does the design achieve stated goals? Conversely, verification refers to whether or not we are properly building and using what we designed (Buede & Miller, 2024). Verification is achieved through submittal reviews, inspections, and rigorous testing methods—such as unit testing, integration testing, factory acceptance testing, site acceptance testing, and other cold commissioning and hot commissioning methods.

Finally, effective requirements management involves continuously maintaining and tracking changes to requirements through regular, period inspections and maintenance. This ensures that any modifications are systematically reviewed, approved, and communicated to all stakeholders. In summary, RE is a critical discipline that fosters clear communication and minimizes misunderstandings by ensuring that requirements are uncovered, cleaned, validated, and implemented properly. RE is the basis for trustworthiness.

Cognitive Work Analysis (CWA)

CWA is an advanced, integrative analysis framework that supports the design and understanding of complex STS by examining the cognitive demands imposed on individuals and groups (Rasmussen et al., 1990; Vicente, 1999). CWA is an evolution of hierarchical task analysis (Stanton, 2006) combined with Rasmussen's Abstract Decomposition Space (Rasmussen & Lind, 1981) and Rasmussen's Decision Ladder (Rasmussen & Goodstein, 1985). Defined by Vicente (1999) as a means to reveal the deep structure of work environments, CWA extends beyond traditional task analysis by considering the broader context of work. CWA has five steps:

1. Work Domain Analysis (WDA) using Rasmussen's Abstraction Hierarchy;
2. Control Task Analysis (CTA) using Rasmussen's Decision Ladder;
3. Strategies Analysis (using Cornelissen's Strategies Abstraction Diagram);
4. Social Organization Structure Analysis; and,
5. Worker Competencies Analysis using Rasmussen's Skills, Rules, Knowledge Framework.

Phase 1. WDA:
WDA is the foundational phase where the overall process is mapped, from high-level purpose to each supporting tool. It involves identifying the purposes, functions, constraints, and resources of the system, thereby providing a high-level understanding of the operational context for performing work.
Phase 2. CTA:
In the CTA phase, the focus shifts to the strategies and decision-making processes that users employ to achieve their objectives within the constraints identified during WDA. This phase examines the cognitive demands of specific tasks and the information and tools required to complete them.
Phase 3. Strategies Analysis:
Strategies Analysis identifies the various methods and approaches that users can adopt to accomplish their tasks, emphasizing flexibility and adaptability in behavior. It explores alternative approaches, allowing for a variety of strategies that can be employed under different conditions.
Phase 4. Social Organization and Cooperation Analysis:
This phase examines the social structures, roles, and interactions that enable effective teamwork and coordination within the system. It is particularly important in environments where collaborative work is essential.
Phase 5. Worker Competencies Analysis:
The final phase focuses on identifying the skills, rules, knowledge, and capabilities required for effective system operation. This analysis ensures that

the design supports the development and utilization of these competencies. Ideally, most tasks can be completed as skill-based or rule-based tasks, because these entail light cognitive loads, and are most comfortable, effective, and efficient for people.

These analysis methods may be compared in order to validate requirements, to identify fake requirements and missing requirements (see Figure 4.3).

For complex, interactive architectural systems designers, it is fortuitous that CWA maps cleanly onto the Systems Modeling Language (SysML), an Object-oriented systems engineering method for architecting complex, interactive systems of systems (Manganelli, 2013). CWA thus bridges the gap between human factors engineering and systems engineering, offering a holistic, user-centered approach that ensures systems are both technically robust and cognitively supportive. In practical terms, applying CWA in the built environment can be leveraged as part of level 3 or level 4 advanced EBD as a foundational method for designing intelligent, adaptive environments—ranging from resilient urban infrastructure to flexible smart buildings—that are tailored to meet the evolving demands of their users while reducing cognitive load, stress, error potential and enhancing overall system efficiency.

Systems Modeling Language

Computer, software, and systems engineers use abstraction diagrams to map the requirements, behaviors, and structures of systems of systems. The SysML is a set of diagrams used for this purpose (see Figure 4.4). SysML has the benefit that it is logically rigorous to the point where multi-objective parametric optimization simulations can be run with the model internally to the modeling tools or using Modelica or some

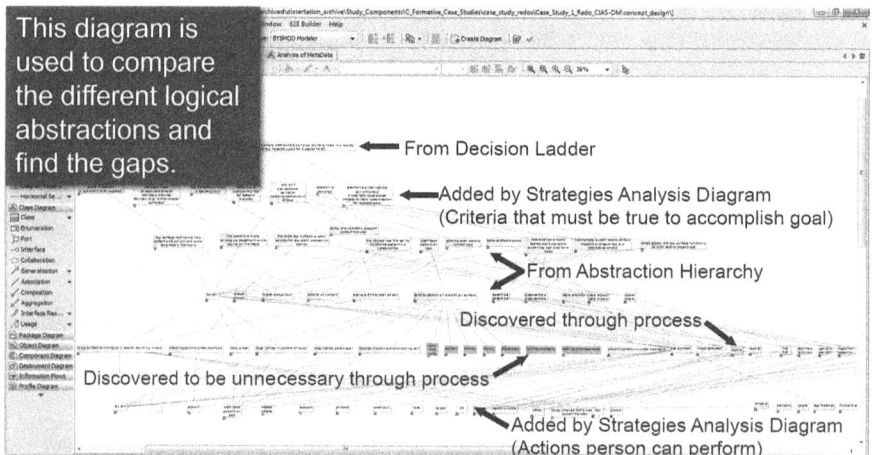

This diagram is used to compare the different logical abstractions and find the gaps.

From Decision Ladder

Added by Strategies Analysis Diagram
(Criteria that must be true to accomplish goal)

From Abstraction Hierarchy

Discovered through process

Discovered to be unnecessary through process

Added by Strategies Analysis Diagram
(Actions person can perform)

FIGURE 4.3 Comparison between the content of the CWA diagrams can help identify system components missing from a design concept, as well as ones that may have initially seemed necessary but are not actually needed (Manganelli & Brooks, 2015).

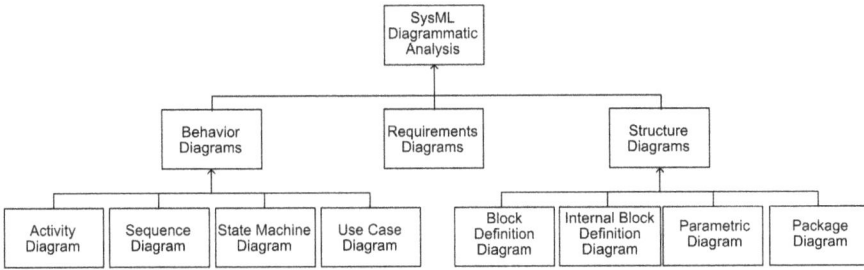

FIGURE 4.4 Designing intelligent, interactive environmental systems will typically require diagramming the requirements, systems, and behaviors in order to capture and address all major, relevant concerns.

other similar simulation platform (Friedenthal et al., 2014). SysML is typically used for large aerospace and defense projects. But there is a strong argument, and some experimental work, that its use must extend to building design in order to realize intelligent, interactive environments (Manganelli, 2013).

Calibrated Trust: Human Trust in Intelligence Systems

Calibrated trust is defined as the appropriate level of confidence that human users place in automation (including AI systems), ensuring reliance is neither excessively high nor unduly low. Calibrated trust has two dimensions: trust in the autonomous system versus the autonomous system's actual capability (which is also referred to as its trustworthiness). According to Lee and See (2004) and McDermott and Brink (2019), trust in automation must align accurately with a system's actual capabilities and limitations. In contemporary society, where AI is becoming pervasive across sectors such as healthcare, finance, autonomous vehicles, and personal digital assistants, calibrated trust becomes crucial for maximizing safety, effectiveness, and user satisfaction (Lee & See, 2004).

Fundamentally, calibrated trust necessitates a clear understanding of an AI system's reliable capabilities and its operational boundaries. Hoff and Bashir (2015) highlight the importance of adequately informing users about AI's strengths and limitations to establish realistic expectations. Overtrust in AI may lead to complacency, resulting in overlooked critical errors or underestimated necessity for human intervention. Conversely, undertrust can prevent full utilization of AI benefits, causing suboptimal decision-making and missed opportunities for enhanced productivity and innovation.

Several empirical factors significantly shape calibrated trust. Following the principles of trustworthy behavior (i.e., human agency and oversight, technical robustness and safety, privacy and data governance, transparency, diversity, non-discrimination and fairness, environmental and societal well-being, and accountability; European Commission, 2019), transparency and explainability, and the degree to which an AI's decision-making process is clearly presented and understandable to humans, are paramount. Dzindolet et al. (2003) emphasize transparency as fostering trust, allowing users insight into how decisions are made. Complementarily, explainability provides detailed rationales behind AI data synthesis and actions, permitting users to verify outcomes and understand internal mechanisms. Reliability also critically influences calibrated trust. Consistent performance and accuracy are essential as AI systems integrate into mission-critical applications. Empirical evidence supports that reliability builds lasting trust since users are likelier to depend on consistently accurate systems (Lee & See, 2004).

User experience further shapes calibrated trust. Mayer et al., (1995) underline that overall user experience—encompassing ease of use, responsiveness, and perceived system empathy—strongly influences trust. Designing AI systems to achieve calibrated trust demands a multidisciplinary approach, integrating human-computer interaction (HCI), cognitive psychology, and ethics. Mayer et al. (1995) stress understanding human trust dimensions alongside technical system performance. User education is crucial; training programs and interactive tutorials significantly enhance user understanding of AI functionalities, limitations, and optimal usage. Adaptive interfaces that modulate complexity based on user proficiency enhance trust calibration by providing advanced users detailed information and simplifying interactions for novices (Parasuraman & Riley, 1997). Continuous feedback mechanisms further advance calibrated trust development.

Ethical considerations are integral to fostering calibrated trust. The IEEE Ethically Aligned Design guidelines (IEEE, 2017) advocate embedding fairness, accountability, and privacy within AI development. Ethical AI adherence to regulatory standards strengthens user trust, ensuring unbiased decisions and secure handling of personal data. This is particularly critical in facial recognition technology, where strict ethical compliance is necessary to prevent misuse and sustain public confidence (IEEE, 2017).

System adaptability further influences calibrated trust. AI systems capable of learning and evolving through real-world interactions must build resilient trust relationships in order to be useful. Adaptive systems employing machine learning continuously update models, responding appropriately to changing conditions.

Calibrated trust also transcends individual interactions, impacting broader societal perspectives. Public trust in AI significantly affects its widespread adoption, influencing regulatory frameworks and market dynamics. Transparent communication regarding AI capabilities and limitations by corporations and governments fosters informed public dialogue and encourages responsible AI deployment. Such societal considerations are paramount in sensitive areas such as autonomous weapon systems or widespread surveillance technologies (European Commission, 2019).

The Validation Square and the Design Science Research Method

The Validation Square (Pedersen et al., 2000) and the Design Science Research Method (Peffers et al., 2007) provide ways for evaluating the efficacy of new systems and technologies by systematic assessments that build confidence that the proposed design intervention is likely *useful with respect to a purpose* and grounded in a deep understanding of the context, existing literature, and available means and methods. The Validation Square assesses proposed new systems designs against *theoretical structural validity* (does the solution align with established knowledge and methods?), *empirical structural validity* (does the solution appear to meet the required design criteria?), *empirical performance validity* (does use in case study projects confirmed targeted results?), and *theoretical performance validity* (do the results achieved and the basis of design suggest that the solution will be able to accommodate application in real use situations in a practically useful (and not just experimental) way?).

Using the Design Science Research Method, a proposed technological solution may result from either: (a) solving a known problem, (b) addressing a known goal, (c) adapting an existing system to a new purpose, or (d) observing a likely useful system. No matter the point of entry for systems development, the developer then proceeds through the stages of: (1) problem definition; (2) outcome objectives definition; (3) design and initial prototyping; (4) iterative, formative testing of working prototype and continued refinement until objectives are achieved; (5) summative testing to confirm performance meets or exceeds objectives on multiple case studies; (6) communication of the solution.

Both methods can be used to validate that a new system is appropriate for its purpose and constructed well and operates well. In addition, the two methods are complementary—and complementary with EBD, as well as STS development methods—and can be combined and modified easily. Lastly, both provide a useful and light-weight framework through which design of systems for which there are not pre-existing mental models can be developed and evaluated.

Enterprise Architecting

Enterprise Architecting is a strategic discipline that aligns an organization's business strategy with its information technology infrastructure to achieve optimal performance, agility, and scalability. According to The Open Group (2018), Enterprise Architecting is the process of designing, planning, and implementing an integrated framework that brings together an organization's structure, processes, information systems, and technologies to work cohesively toward common business objectives. This holistic approach is not only about reducing redundancies and streamlining operations; it is also about ensuring that every technological investment is directly linked to strategic initiatives, thereby enhancing decision-making, operational efficiency, and innovation.

The discipline is guided by several well-established frameworks, including The Open Group Architecture Framework (TOGAF), the Zachman Framework, and the Department of Defense Architecture Framework (DoDAF) (Dimitrov, 2012), all of which offer structured approaches for organizing and managing architectural artifacts. TOGAF, for instance, emphasizes a cyclical process of planning, design, implementation, and governance, ensuring that the architecture evolves in response to new challenges (The Open Group, 2018). The Zachman Framework, on the other hand, provides a multidimensional schema that categorizes enterprise artifacts from multiple perspectives, facilitating a clear and methodical understanding of organizational components (Zachman, 1987). DoDAF incorporates many useful "views," such as the Operational View 1 (OV-1) diagram shown in Figure 4.5. The OV-1 shows a system of concern nested within the larger system of systems within which it operates.

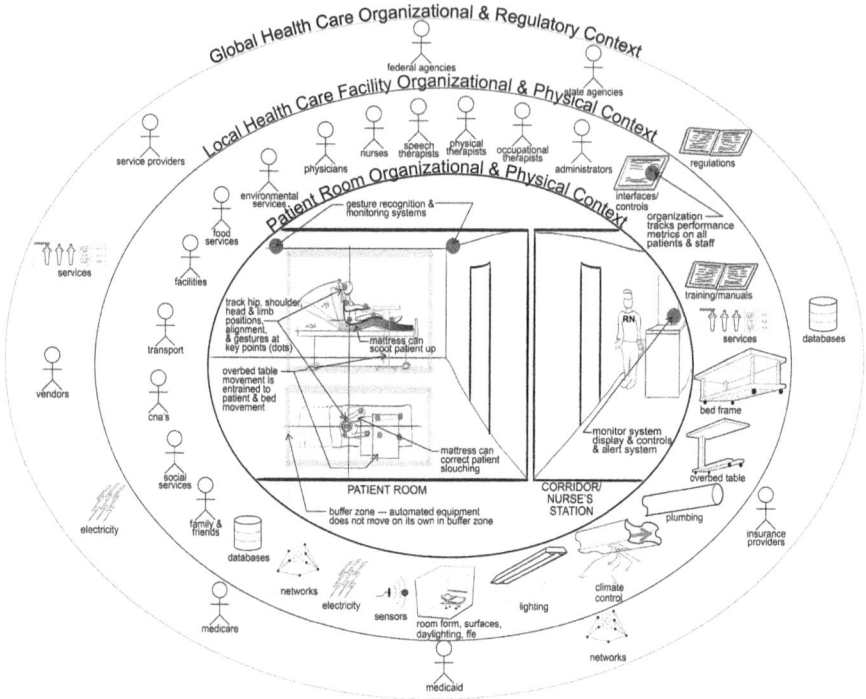

FIGURE 4.5 A DoDAF OV-1 diagram mapping the system of concerns that must be addressed to design, construct, and operate an intelligent, adaptive hospital patient room (Manganelli, 2013).

Key components of Enterprise Architecting include:

Component 1. Business Architecture:
Business Architecture defines the organization's strategic vision, governance structures, organizational design, and key business processes.
Component 2. Information Architecture:
Information Architecture structures the organization's data assets and delineates the flows of information within the enterprise.
Component 3. Application Architecture:
Application Architecture outlines the individual software applications and describes how they interact within the enterprise ecosystem.
Component 4. Technology Architecture:
Technology Architecture specifies the hardware, software, and network infrastructure required to support the organization's operations.

RAY IV: EMBODIED ACTION AND COGNITION

Neuroergonomics and Neuroadaptive Systems

Neuroergonomics is defined as the interdisciplinary study of the human brain in relation to performance in work and everyday settings, integrating principles from neuroscience, ergonomics, and human factors engineering (Parasuraman & Rizzo, 2009). It seeks to understand how brain functions such as attention, memory, decision-making, and motor control interact with work systems to optimize human-system interactions, thus enhancing safety, efficiency, and overall well-being (Parasuraman & Rizzo, 2009). Furthermore, neuroergonomics significantly aids in mitigating fatigue and stress. Research demonstrates that by monitoring neural markers indicative of cognitive overload, adaptive systems can adjust environmental factors such as lighting or ambient sound to maintain optimal cognitive performance (Fairclough & Gilleade, 2014).

Neuroadaptive systems are a closely related set of technologies to neuroergonomics. Specifically, neuroadaptive systems are able to interpret human behavior in order to assess human system states and adapt workload and/or work processes to enhance human performance and/or well-being (Hettinger et al., 2003). For instance, consider a worker on an assembly line. When the work starts and the worker is refreshed and alert, the system does not help much, instead merely monitoring human performance. This allows the human to develop or maintain skills, and to develop resilience, stamina, and flexibility. As the worker's performance degrades due to fatigue, the system may selectively modify work and environmental parameters, such as conveyor speed, lighting, temperature, air change rate, in order to help the worker maintain performance. When indirect task adjustments no longer help the worker maintain performance, then the system may gradually assume more of the workload itself, so that the overall pace of production remains the same. In this last situation, the roles switch, to a degree, and the human's role becomes that of assistant and monitor as the neuroadaptive system performs more of the work. In this way, both human and machine learn from each other, practice and self-optimize individually and as a team, together, while also each co-developing problem-solving and adaptation capacities.

Embodied Cognition (Including Direct and Indirect Perception)

Embodied cognition is defined as a theoretical framework proposing that cognitive processes are deeply rooted in bodily interactions with the environment, asserting that cognition is inseparable from sensory and motor functions (Lakoff & Johnson, 1999). Wilson (2002) summarizes six common meanings behind the term, "embodied cognition":

1. Cognition is situated;
2. Cognition is time-pressured;
3. Cognition is offloaded onto the environment (i.e., co-processed with the environment);
4. The environment is part of the cognitive system;
5. Cognition is for action; and,
6. Offline cognition (i.e., in the head) is still body-based (i.e., based on how we engage the world as an extension of our consciousness).

Expanding on this idea, Clark (2003) notes that,

In all this we discern two distinct, but deeply interanimated, ways in which biological cognition leans on cultural and environmental structures. One way involves a developmental loop, in which exposure to external symbols adds something to the brain's own inner toolkit. The other involves a persisting loop, in which ongoing neural activity becomes geared to the presence of specific external tools and media.

Expanding on this, Dourish (2001) emphasizes that interactions with technology are inherently embodied, situated within physical, social, and cultural contexts, profoundly influencing perception, action, and thought.

Direct perception, within the subdiscipline of ecological psychology, as introduced by Gibson (1979), posits that individuals immediately recognize and use *affordances* in their environment, such as perceiving a rock as "throwable" versus "step-able" versus "sit-able," without requiring complex internal processing. Conversely, indirect perception involves higher-level cognitive processing, integrating sensory inputs into mental representations, as seen when reading and interpreting maps. But it is more than this. There are many apparently higher-order cognitive processes that are actually simpler interactions between animals and their environments. For instance, a baseball player is not doing calculations in his head and then running and catching a fly ball. Rather, the baseball player's visual system is able to quickly assess and maintain an angle of incidence between the player and the ball in order to track to the ball, without requiring much, if any higher cognition (see Figure 4.6).

Why does this matter? It matters because humans are incredibly sensitive to changes in their environments and tools. How sensitive? As per below, just 40 hours practicing golf by middle-aged people generates noticeable changes in gray matter in the brain. How much more powerful, then, is the impact of one's environment, where one lives and works for thousands of hours per year? Moreover, and critically, this remapping occurs very fast, beginning within minutes of tool use and resulting in stable mappings within days to months for a single skill.

Iriki, Tanaka, and Iwamura (1996) note:

A tool is an extension of the hand in both a physical and a perceptual sense. The presence of body schemata has been postulated as the basis of the perceptual assimilation of tool and hand. We trained macaque monkeys to retrieve distant objects using a rake, and neuronal activity was recorded in the caudal postcentral gyrus where the somatosensory and visual signals converge. There we found a large number of bimodal neurons which appeared to code the schema of the hand. During tool use, their visual receptive fields were altered to include the entire length of the rake or to cover the expanded accessible space. These findings may represent neural correlates of the modified schema of the hand in which the tool was incorporated.

The results of the other new study suggest that, when monkeys learn a challenging tool-use task, they grow new connections in their brains. Hihara et al. trained adult monkeys to use a rake to retrieve pieces of food. Their earlier work showed that, after three weeks training, monkeys treat the tool as an extension of their own body. This phenomenon, which is also observed in humans using tools, can be seen both behaviourally and by recording from neurons in the parietal cortex.

(Johansen-Berg, 2007)

While this was demonstrated in monkeys, the same effect has been found to take place in humans,

Physical
Information
Processing

Non-phyisical
Information
Processing

Direct
Perception

Logical Information Processing Continuum

Indirect
Perception

Hardware, e.g.
- Rocks
- Mechanical clock
- Keys/locks
- Columns & beams
- Utensils

Software/Hardware Integrations
- Biological reflexes
- Anti-lock braking
- Team sports
- Visually tracking a baseball in flight
- Assembly lines

Software, e.g.
- Abstract analogical thought
- Concept models
- Computer simulations

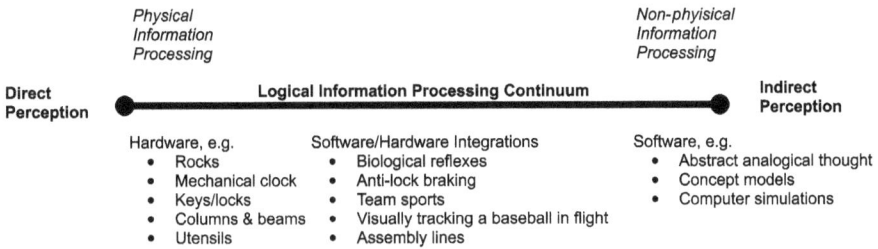

FIGURE 4.6 The spectrum of information processing from direct perception to indirect perception, in which all information processing is "thinking"—even physical acts (Manganelli, 2018).

Previous neuro-imaging studies in the field of motor learning have shown that learning a new skill induces specific changes of neural gray and white matter in human brain areas necessary to control the practiced task..... In the present longitudinal MRI study, we used voxel-based morphometry to investigate training-induced gray matter changes in golf novices between the age of 40 and 60 years, an age period when an active lifestyle is assumed to counteract cognitive decline. As a main result, we demonstrate that 40h of golf practice, performed as a leisure activity with highly individual training protocols, are associated with gray matter increases in a task-relevant cortical network encompassing sensorimotor regions and areas belonging to the dorsal stream..... Thus, we demonstrate that a physical leisure activity induces training-dependent changes in gray matter and assume that a strict and controlled training protocol is not mandatory for training-induced adaptations of gray matter.

(Bezzola et al., 2011)

Skills, Rules, and Knowledge Framework

Rasmussen recognized that human activity can be classified into the categories of skill-based behavior, rule-based behavior, and knowledge-based behavior (Rasmussen, 1983). Skills are responses that can be processed as a response to a signal with little to no cognition and little to no rule-following. Reflexes are examples of skills. Similarly, moving a computer mouse is an example of a skill. These are the most time-efficient and energy-efficient cognitive behaviors because there is little to no active cognition required. Rule-based behaviors are the second most economical types of behaviors. Rule-based behaviors are mostly skill-based behaviors that require a bit of recognition and/or recall and then binary or branching logic decision-making. Following rules and recognizing familiar patterns require a small amount of active cognition but are very efficient processes. Most behaviors are rule-based behaviors. Knowledge-based behaviors are the most resource-intensive behaviors cognitively. Knowledge-based behaviors include pattern formation activities, rule creation activities, skill formation activities, and planning activities.

When designing ULSS and HART, it is useful to classify user behaviors—for both biotic and abiotic agents—according to the skills, rules, and knowledge framework. Moreover, it is best to make agent actions skill-based or rule-based to the greatest extent possible. This makes the system overall less error prone and more sustainable, efficient, and resilient for the users.

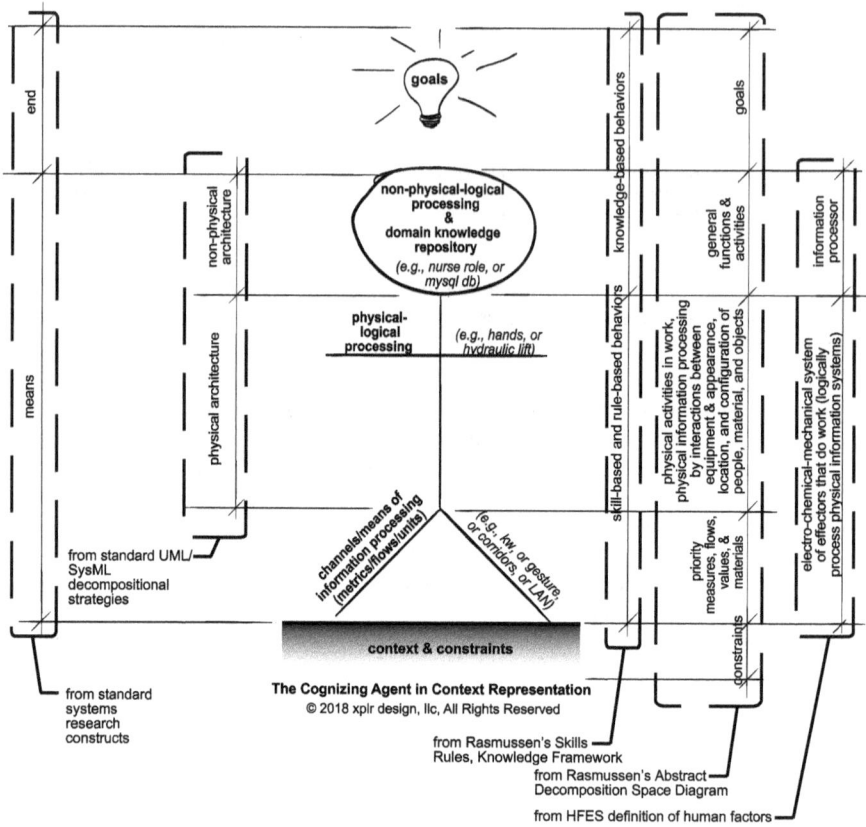

FIGURE 4.7 Integrated means/end hierarchy diagram, integrating frameworks from systems research, human factors, systems engineering, and Rasmussen's Abstraction Hierarchy (stage 1 of CWA) (Manganelli, 2018).

All of the above information is summarized and integrated in the reference diagram presented in Figure 4.7.

...HUMAN-IN-THE-LOOP

RAY V: HUMAN-IN-THE-LOOP AUTOMATED SYSTEMS

Designing Environments to Support Human Cognition

Designing environments that support human cognition and action starts with acknowledging that designing the environments in which people live and work is designing parts of their sensing, their perception, their cognition, and their actions. It involves creating spatial conditions that enhance mental processing, learning, decision-making, and overall well-being. An interdisciplinary approach is required, drawing on cognitive science, human factors engineering, systems engineering, psychology, architecture, HCI, software development, and many other professional perspectives. It aims to *cultivate environments* (in the ULSS/STS sense) where physical and digital realms are seamlessly integrated in support of human information processing. Contemporary design must transcend rigid forms and accommodate fluctuating patterns of human behavior and technological interaction (Kirsh, 2013). The Design Project Ecological Niche Construction Checklist provides a conceptual framework for designing to support cognition and action (Manganelli, 2016; see Figure 4.8). It integrates Kirsh's concepts of activity spaces and performance design with Rasmussen's Skills, Rules, and Knowledge Framework to help designers categorize the activities that take place within a space based on the intended performance targets for those activities, as well as the skills, rules, and knowledge loads they require of the agents, and the information systems, tools, and environmental affordances provided to support the activities.

In summary, designing environments that support human cognition is a multifaceted endeavor that requires the integration of insights from cognitive science,

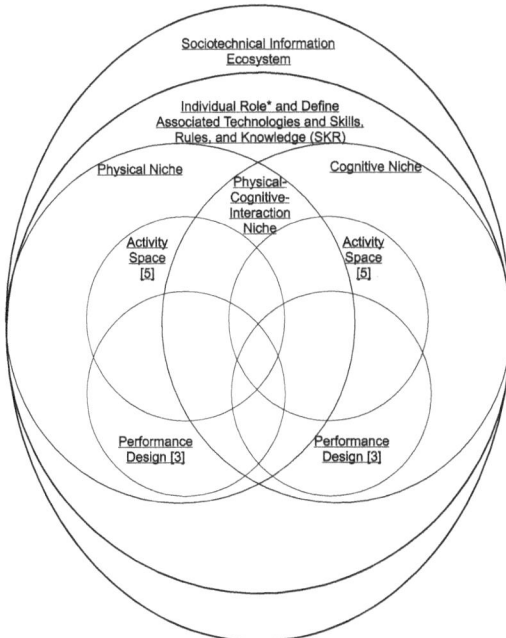

FIGURE 4.8   The Design Project Ecological Niche Construction Checklist helps designers to consider the aspects of system design that support human action and cognition (Manganelli, 2016).

cognitive psychology, architecture, and HCI. Drawing on seminal works by Kirsh, Rasmussen, and others, it is clear that adaptive, user-centered, and technologically integrated designs can significantly enhance cognitive performance and reduce cognitive load. These environments not only facilitate better decision-making and creative problem-solving but also contribute to the overall satisfaction and well-being of their users.

Human-Environment Information Co-processing: Enhancing Intelligence Through Integration

In practical terms, cognitive scaffolding can be observed in various contexts such as healthcare, education, transportation, and urban planning. Healthcare environments that integrate intuitive wayfinding, clear signage, and calming aesthetic elements demonstrably reduce patient anxiety, enhance cognitive clarity, and improve overall patient and staff outcomes (Ulrich et al., 2008). Similarly, educational environments leveraging digital integration—such as interactive whiteboards, tablets, and real-time feedback systems—significantly enhance cognitive engagement, knowledge retention, and learner satisfaction (Wu et al., 2013).

In conclusion, human-environment information co-processing offers a robust conceptual and empirical framework for understanding cognitive processes as fundamentally distributed across the brain, body, and external environment. The future evolution of human-environment information co-processing necessitates interdisciplinary collaboration among architects, engineers, designers, cognitive scientists, technology developers, ethicists, and policymakers. Emerging technologies, such as pervasive computing, AI, and immersive virtual environments, promise unprecedented opportunities for cognitive enhancement, provided they are responsibly and ethically integrated (European Commission, 2019). By aligning design practices with established principles from cognitive science, ergonomics, HCI, and ethical frameworks, future environments can more effectively augment human cognition, productivity, and well-being, while mitigating potential ethical and social risks.

Ultimately, acknowledging and leveraging the distributed nature of human cognition through carefully designed human-AI-environment interactions will be critical for enhancing intelligence and navigating an increasingly complex, digitally mediated world.

Human-in-the-Loop/On-the-Loop Problem-Solving with Ultra-Large-Scale Systems

Within ULSS, human operators occupy a pivotal "in-the-loop" position, and human monitors occupy a pivotal "on-the-loop" position, actively engaging in monitoring system states, decision-making processes, and taking corrective actions essential for efficient and safe operations (Parasuraman & Wickens, 2008). Human-in-the-loop/human-on-the-loop problem-solving capitalizes on the complementary strengths of both humans and machines. Humans contribute vital cognitive capacities, including creativity, intuition, ethical judgment, and nuanced decision-making, while machines provide the advantages of rapid data processing, high precision, and effective management of extensive information streams (Parasuraman et al., 2000). In healthcare contexts, for instance, ULSS technologies can deliver real-time analytics and diagnostic support, enhancing physicians' clinical decision-making capabilities while preserving ultimate human responsibility for final diagnostic and treatment decisions. Research in human factors underscores the importance of maintaining human oversight to address patient-specific complexities and ethical considerations that automated systems alone may not adequately handle (Karsh et al., 2010). Similarly, in autonomous transportation systems, real-time environmental and vehicle data analyses continuously optimize routing and enhance safety measures, but human oversight remains crucial, particularly in scenarios involving unexpected environmental variations or system anomalies (Endsley, 2017).

Enhancing Intelligence Through Integration

The integration of human-in-the-loop and on-the-loop problem-solving within ULSS presents profound implications beyond merely optimizing technical performance. As digital and physical processes increasingly intertwine, distinctions between human cognition and technological augmentation become progressively blurred, extending the scope and depth of human cognitive capabilities. Clark's (2008) framing of human-environment co-processing describes how human intelligence is continuously augmented by seamlessly integrated technological supports. This synergistic relationship not only amplifies problem-solving capabilities but fosters more adaptive, resilient, and responsive cognitive systems (Clark, 2008; Smart, 2017).

Addressing ethical and societal implications of increasingly integrated ULSS is equally vital. Designing CPS to complement human cognition without eroding autonomy or infringing upon privacy is a substantial contemporary challenge. Comprehensive ethical frameworks, such as those articulated in the IEEE Ethically Aligned Design guidelines and the European Commission's Ethics Guidelines for Trustworthy AI, advocate for principles like transparency, accountability, and user-centered governance to safeguard individual rights and societal values (IEEE, 2017; European Commission, 2019). As these technologies gain broader societal integration, maintaining public trust and establishing robust regulatory oversight become indispensable, ensuring that technological advancements consistently serve human interests and well-being (Turkle, 2011; European Commission, 2019).

...SYSTEMS

Ray VI: Human-Built Environmental Systems

Trans-Programming: Bridging Dimensions

Trans-Programming redefines the architectural program as a dynamic, evolving field wherein spatial narratives and unfolding events continuously reshape the intended use. Rather than perceiving a building's functions as a fixed checklist, this perspective conceptualizes architectural spaces as dynamic "fields of potentialities," realized through continuous interactions among design intentions, user experiences, and contextual influences (Tschumi, 1996, 2001). In this paradigm, the architectural program transcends a static list, emerging instead as a fluid and responsive structure open to continuous reinterpretation.

Central to Trans-Programming is the premise that architecture unfolds through events rather than merely implementing a predetermined brief. Bernard Tschumi (1996) illustrates this concept vividly through his example of an inhabited bridge—a structural form intentionally designed for flexible occupation, accommodating various activities such as exhibitions, public gatherings, or temporary performances. According to Tschumi (1996), architecture fundamentally "is as much about the event that takes place as about the space itself." This provocative stance disrupts the traditional "form follows function" philosophy, advocating instead for designs rich in latent possibilities awaiting activation through user interactions and evolving contexts.

Trans-Programming also employs the principle of disjunction—a deliberate disruption of traditional spatial orders—to produce multiple interpretive possibilities simultaneously. Tschumi (2012) describes disjunction not as a design flaw but as an intentional strategy enabling numerous concurrent uses within a singular architectural environment.

The theoretical underpinning of Trans-Programming draws heavily from post-structuralist ideas. For example, architectural theorist Louis Martin (1990) argues for interpreting architecture as a textual construct, comprising signs and symbols open to multiple readings rather than fixed meanings. Within this theoretical lens, the architectural program itself becomes a dynamic text, constantly rewritten through everyday human interactions. Such iterative processes resonate with adaptive design theories proposed by Alexander (1977), who contends that environments capable of ongoing feedback and adaptability inherently achieve greater resilience and responsiveness.

Architects embracing Trans-Programming frequently deploy design strategies that cultivate flexibility and continuous spatial reconfiguration. Modular and movable design elements exemplify this approach, allowing easy rearrangement and adaptability to shifting user requirements. An iterative and reflective design methodology is vital for Trans-Programming. Instead of adhering strictly to static programs, architects must foster continuous dialogues with building occupants. Tschumi (2001) emphasizes that "architecture should constitute a dialogue between designer intentions and user appropriations," allowing user experiences and feedback to iteratively refine spatial configurations.

Further grounding Trans-Programming, Norman's (2013) empirical findings highlight the cognitive benefits of spatial adaptability. Environments intentionally designed for continuous reconfiguration actively engage occupants, inviting creative participation in reshaping their surroundings, significantly enhancing satisfaction and promoting innovative space use. Additionally, participatory design methodologies advocated by Sanders and Stappers (2008) show that involving users directly in spatial adaptations generates meaningful environmental interactions, deepening the connection between users and spaces—a fundamental principle of Trans-Programming. Real-world applications demonstrate these insights across diverse contexts, from adaptable exhibition spaces to innovative residential interiors and community hubs.

In conclusion, Trans-Programming fundamentally redefines architectural programming as a continuous dialogue between spatial forms and dynamic events. It is complementary to STS design processes, the design processes of ULSS, and therefore also well-suited to be part of the method of designing the built environment to incorporate Industry 6.0 technology. Trans-Programming establishes a transformative framework, empowering architects to craft environments capable of continuous adaptation, resilience, and multiplicity—qualities essential to sustaining meaningful human experiences within rapidly changing contemporary contexts.

Cross-Programming: Synergizing Systems

Cross-Programming is an architectural strategy that synthesizes multiple programmatic functions within a unified spatial framework, challenging the conventional practice of segregating uses. Instead of treating functions as isolated entities, Cross-Programming envisions them as interrelated components that can overlap, interact, and mutually reinforce one another. Tschumi (2012) notes that "to cross the program is to liberate it from its own determinism," underscoring the potential for functions to transcend their traditional boundaries when allowed to intersect.

The theoretical foundation for Cross-Programming rests on the idea that the architectural program should be understood as a network of interconnected possibilities rather than a fixed list of requirements. Martin (1990) argues that architecture must dissolve traditional hierarchies, allowing diverse functions to coexist on an equal footing. In Cross-Programming, the program is not compartmentalized; instead, functions are interwoven into a continuous spatial tapestry.

Practical applications of Cross-Programming are evident in projects where flexible spatial zones allow for the simultaneous operation of varied functions. For example, consider a cultural facility designed to host both educational workshops and interactive exhibitions. In a traditional design, these functions might be allocated to separate rooms with rigid boundaries. In contrast, a Cross-Programming approach would integrate these functions within an open, fluid space, utilizing modular partitions and shared circulation routes to facilitate interaction. This dynamic reconfiguration is achieved through iterative design processes, as Tschumi (2012) emphasizes that the built environment must be in a constant state of dialogue with its users. This mediatory role is further reinforced by cybernetic theories, which argue that systems designed with feedback loops are inherently more capable of adapting to change (Wiener, 1961).

Super-Imposition: Layering Realities

Super-Imposition is an architectural strategy that systematically layers multiple programmatic elements within a single spatial framework to create a composite, multidimensional narrative. Rather than compartmentalizing functions into distinct zones, this approach envisions a building as a stratified text—each layer contributing unique, overlapping potentials that enrich the overall experience. Tschumi (2012) emphasized that "the program is not a fixed container but a series of overlapping potentials," a concept that forms the basis for Super-Imposition and has been further elaborated in his subsequent writings (Tschumi, 1996, 2005).

Fundamentally, Super-Imposition challenges traditional spatial organization by advocating for a design in which functions coexist in layered, interpenetrating zones.

In this paradigm, the architectural program is not divided by clear-cut boundaries; instead, it is arranged in overlapping strata that allow for multiple, simultaneous interpretations. The methodology uses staggered floor levels, mezzanines, and flexible partitions to create semi-transparent layers that allow light, sound, and movement to flow between different zones. These strategies ensure that while each layer maintains a degree of autonomy, it also contributes to a larger, cohesive whole. Tschumi (2005) posits that "to layer is to allow space to speak in multiple voices," suggesting that ambiguity and multiplicity are integral to creating a rich spatial experience. This notion challenges the conventional pursuit of clarity and singular purpose, instead celebrating complexity as a source of creative potential.

The role of the architect in Super-Imposition is that of a facilitator who orchestrates the interplay between multiple spatial layers. Rather than dictating a fixed use for each zone, the architect designs for flexibility, ensuring that each layer can be reinterpreted as conditions change. This iterative process of design and reconfiguration is supported by adaptive design theories (Alexander, 1979) and cybernetic models (Wiener, 1961), which emphasize that systems capable of continuous adjustment are inherently more robust (Table 4.1).

TABLE 4.1
Architectural Strategies Analysis for Suggested Ways to Designing Adaptable Environments

Criterion	Trans-Programming	Cross-Programming	Super-Imposition
Definition	Reimagines architectural programs as dynamic fields of possibilities where spaces evolve through changing events and user interactions (Tschumi, 2012).	Integrates multiple distinct functions into unified spaces, enabling functions to overlap, interact, and mutually reinforce (Tschumi, 2012).	Systematically layers multiple programmatic elements within one spatial framework, creating multidimensional, overlapping experiences (Tschumi, 2012).
Theoretical Foundation	Post-structuralist theory: program as evolving text continually rewritten by user experience (Martin, 1990; Tschumi, 2012).	Complexity and network theory: functions interrelated in a dynamic spatial network rather than isolated units (Martin, 1990; Alexander, 1979; Tschumi, 2012).	Layered and semi-transparent spatial narratives: embracing ambiguity, complexity, and multiple simultaneous interpretations (Alexander, 1979; Wiener, 1961; Tschumi, 2012).
Methodological Approach	Iterative, event-driven spatial design using flexible architectural elements such as movable partitions and adaptable lighting to facilitate evolving uses (Norman, 2013; Sanders & Stappers, 2008).	Overlapping zones and modular spatial configurations, employing flexible plans, adaptable furniture, and integrated circulation paths to enable simultaneous multi-functionality (Norman, 2013; Sanders & Stappers, 2008).	Stratified spatial design using staggered floors, mezzanines, adjustable partitions, and permeable boundaries to create layered, interpenetrating spatial experiences (Norman, 2013; Sanders & Stappers, 2008).

(Continued)

TABLE 4.1 (*Continued*)
Architectural Strategies Analysis for Suggested Ways to Designing Adaptable Environments

Criterion	Trans-Programming	Cross-Programming	Super-Imposition
Role of the Architect	Facilitator of spatial dialogue: architect as collaborator engaging users directly in continuous reconfiguration and reinterpretation of spaces (Sanders & Stappers, 2008; Tschumi, 2012).	Mediator and integrator of diverse functions: architect orchestrates interactions and overlaps between multiple simultaneous programmatic needs (Sanders & Stappers, 2008; Tschumi, 2012).	Composer of layered experiences: architect strategically organizes overlapping programmatic layers, fostering continual reinterpretation and dynamic interactions (Sanders & Stappers, 2008; Tschumi, 2012).
Adaptability and Flexibility	High adaptability: spaces constantly change based on event-driven user interactions, designed explicitly for long-term dynamic evolution (Norman, 2013).	Functional flexibility: spaces continuously adapt and integrate multiple simultaneous activities, reducing the need for spatial compartmentalization and frequent renovations (Norman, 2013).	Structural flexibility through layering: spatial adaptability enables continuous reconfiguration and sustained relevance without extensive redesign (Norman, 2013).
In Simple Terms	Spaces continuously evolve based on changing events and user activities, with no fixed uses but adaptable scenarios.	Different functions are intentionally mixed and overlapped in the same space, allowing them to support and enhance each other.	Multiple layers of activities coexist within the same area, creating rich and overlapping experiences simultaneously.

Ray VII: Non-Humanoid Social Robotics (NH-SR): Redefining Interaction

NH-SR refers to robotic systems embedded in the built environment—within walls, furniture, installations, or architectural elements—that engage humans in social or collaborative interactions without resembling the human form. Unlike humanoid or toy-scale robots, these architectural-scale systems blur the line between "robot" and "environment," transforming rooms and structures into interactive partners (Mokhtar, 2019). By operating at the scale of architecture, NH-SR systems redefine human-technology interaction: the building itself becomes a social agent that can sense, respond, and adapt to its occupants. This approach leverages the ubiquity of architectural space to integrate robotics "everywhere" in daily life, an idea aligned with Mark Weiser's vision of ubiquitous computing and "calm technology" (Weiser & Brown, 1997). Indeed, Mokhtar (2019) argues that moving beyond humanoid embodiments can help overcome issues like the uncanny valley and limited functional integration, enabling social robots to become a seamless part of our living environments. In essence, NH-SR systems treat interactive architecture as a social interface—walls, floors, and furniture endowed with sensors, actuators, and AI that collectively support human activities, communication, and well-being (Manganelli, 2015; Mokhtar, 2011; Mokhtar, 2025) (Figure 4.9).

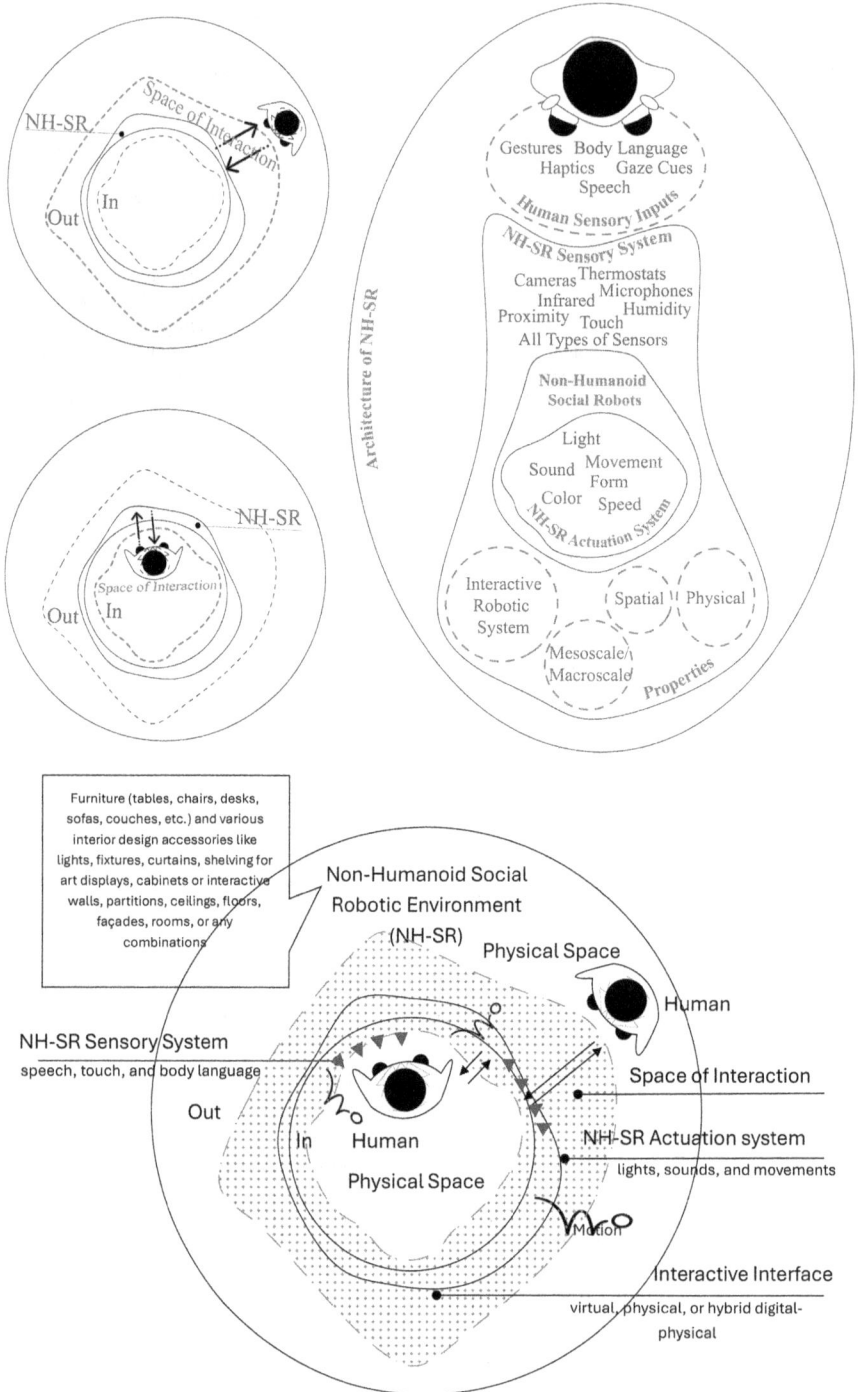

FIGURE 4.9 Human interaction with NH-SRs from outside the robot's environment (top left) or inside the robot's environment (under top left); and (right) the architecture of NH-SRs. (Mokhtar, 2019). The Architecture of NH-SR Environments (bottom) (Mokhtar, 2025).

Educational settings offer fertile ground for NH-SR, as classrooms, laboratories, and museums can be transformed into interactive learning partners. An intelligent classroom environment might feature responsive walls and furniture that actively participate in teaching—for example, walls that display interactive content when students ask questions, or desks that reconfigure for group work versus individual study. Early prototypes such as Mokhtar's intelligent Classroom Environment demonstrate how embedded robotics can adjust lighting, sound, and spatial configuration to improve student engagement and communication (Mokhtar et al., 2018). By sensing student behavior (e.g., movement, noise level, or biometric signals of attention), an educational NH-SR space can adapt in real time—dimming lights and lowering noise partitions when students need focus, or projecting stimulating visuals when energy wanes—effectively acting as an "empathetic" environment tuned to learning rhythms.

Interactive museum and exhibit spaces also exemplify educational NH-SR. A landmark example is "Ada—the Intelligent Space," an installation at the Swiss Expo 2002 that served as a playful learning environment for science and art. Ada was essentially a large room embedded with a dense mesh of sensors and actuators—pressure-sensitive floors, microphones, lights, and speakers—allowing it to detect visitors' movements and respond with orchestrated light and sound patterns (Eng et al., 2003). Dozens of people could interact with Ada simultaneously; as they walked or made noise, the environment reacted in real time, "learning" from their behavior. Visitors described the space as if it were a living creature, and indeed Ada was designed as an artificial creature with its own "mood," capable of engaging users in an open-ended, exploratory dialogue (Eng et al., 2003).

In sum, NH-SR in educational contexts shifts the paradigm from teacher-student or guide-visitor interactions to a triadic relationship including the environment. These smart environments provide adaptive feedback, guiding learners through discovery. They are also culturally adaptable; for instance, an interactive learning wall in a history museum could change its storytelling style based on the cultural background of visitors (detected via language or interactive choices), making education more inclusive.

In homes and residential buildings, NH-SR focuses on personalization, comfort, and assistive living, embedding intelligence into the fabric of domestic space. The goal is an environment that not only automates tasks but also interacts with inhabitants in a social and supportive manner.

A pioneering project in this realm is the Reconfigurable House by Haque and Somlai-Fischer (2007), an experimental full-scale environment constructed from thousands of low-tech sensors and actuators that residents could rewire and reprogram at will. This prototype house, first installed as a public exhibit in Tokyo, was a direct challenge to conventional smart home ideology (Haque & Somlai-Fischer, 2007). Every element of the house was interactive: there were light-emitting "bricks" that meowed like cats when touched, ceiling fixtures with toy penguins that blinked, and LED flowers that changed color when stroked (Vale, 2008). Crucially, any sensor could be connected to any actuator through a simple touchscreen interface, meaning the occupants themselves defined how the house behaved—one could link the floor pressure sensor to the window blinds or make the lamp flash when the sofa was sat on. Because of this open configurability, the Reconfigurable House could develop completely new behaviors over time as its users experimented and adapted it to their preferences (Borg, 2011). Social interaction in this context takes the form of creative dialogue between the inhabitants and the home: the home encourages inhabitants to tinker and, in turn, surprises them with emergent behaviors. Notably, the house was programmed with a bit of personality—if left alone with no user interaction, it would grow "bored" and begin to daydream by autonomously re-networking its devices, thus playfully inviting the humans to re-engage (Borg, 2011).

Beyond experimental projects, commercial applications of residential NH-SR are emerging. Modern apartments are beginning to include robotic furniture that physically transforms to maximize space and utility while responding to residents' commands. For example, the Ori system (commercialized in 2018 from MIT Media Lab research) is a modular furniture unit that glides on tracks to reconfigure a studio apartment into a bedroom at night, a home office by day, or an open living area for social gatherings (Matheson, 2018). The social robotics element comes into play as these furnishings respond fluidly to human schedules and even anticipate needs (e.g., integrating with a smart assistant to prepare the home office setup when a calendar meeting is about to start). In doing so, the furniture behaves as an attentive helper embedded in the architecture.

Residential NH-SR can also focus on emotional and health support. Consider a wall installation that changes its ambient lighting color and pattern to enhance the household's mood—perhaps glowing calm blue hues when occupants are stressed, or dynamic warm patterns when lively conversation is detected. Mokhtar's Belonging Robot, for instance, was a hybrid physical-digital wall system designed to reflect the collective mood of students and staff in a university or university dormitory, promoting a sense of community and emotional connectedness (Mokhtar & Mansour, 2016).

In essence, NH-SR in residences transforms the private realm into an interactive habitat that adapts physically, functionally, and socially. It supports environmental adaptation by adjusting layouts, climate, and resources to personal habits (closing windows when it's noisy outside and the occupant is resting, for example). It augments social interaction by serving as an active participant in daily routines—whether playfully cooperating in customization as in the Reconfigurable House, or providing comfort and companionship by "understanding" the inhabitants. This leads to a richer user experience of home: a place that is responsive and alive, tailoring itself to fit each family's life patterns and strengthening the bond between people and their living spaces.

Public installations and architectural art projects have been at the forefront of NH-SR, often aiming to inspire wonder, provoke thought, or facilitate community interaction in plazas, parks, and cultural exhibitions. In these large-scale deployments, the emphasis is on creating shared social experiences and responsive environments that engage diverse audiences. Architectural-scale robots in public spaces must be robust and intuitive, and they often draw from biomimicry and kinetic art to resonate with people on an emotional level (Beesley, 2010; Baumeister, 2014).

One celebrated example is "Hylozoic Ground" by architect Philip Beesley, which premiered as Canada's pavilion at the 2010 Venice Architecture Biennale. Hylozoic Ground is an immersive, interactive architectural environment constructed from tens of thousands of lightweight components that behave like a living forest canopy

(Beesley, 2010). Integrated sensors detect the presence and touch of visitors walking through the installation, and then shape-memory alloy actuators trigger waves of motion—panels flutter open and closed, fronds reach out or retract. The entire environment gently "breathes" with diffuse pulsing patterns of motion and light, as if it were a surreal synthetic organism. The user experience is deeply contemplative and emotional; many find it calming or dream-like to wander through, suggesting that such NH-SR installations can provide moments of reflection and connection in otherwise impersonal public venues.

Another important category of public NH-SR focuses on civic interaction and cultural expression. Monumental-IT, a project developed by Mokhtar (2011) as part of his doctoral work of Clemson University supervised by Green, Walker, and Lauria, is an example of an intelligent robotic monument designed to reflect the voices of the community. In this concept (prototyped as an interactive installation), the structure of the monument can change shape or display patterns in response to input from lay citizens—for instance, people could submit messages or emotions via mobile phone, and the monument would physically reconfigure or light up to represent these collective sentiments (Mokhtar et al., 2013). This turns a traditionally static civic symbol into a living communicative entity, effectively "giving form to the voices of lay-citizens." The social robot here is the monument itself, which engages the public in dialogue: it listens (takes input) and then responds in a way that is visible and tangible to everyone around.

Public urban spaces have also seen more whimsical NH-SR installations aimed at delight and social bonding. For example, Daan Roosegaarde's "Dune" (2007–2011) is a series of interactive landscapes along pedestrian pathways, composed of hundreds of fiber-optic reeds that react to the sounds and movements of passersby. As people walk by at night, the tall reeds light up and rustle in waves, essentially conversing with the pedestrians through light and sound (Roosegaarde, 2011). Children often start running and laughing to make the entire line of reeds come alive, and strangers find themselves smiling at each other, a shared moment created by the environment.

It's important to note how architectural integration is handled in public NH-SR. These systems are often designed to be visually compelling even when dormant, merging art with function, so that they contribute positively to the cityscape. They use materials and forms that resonate with human perception (soft fabrics, natural movement patterns, familiar symbols) to ensure approachability. This helps diverse public users—young, old, tech-savvy or not—to intuitively engage. Thus, the city itself becomes a facilitator of social connection. Environmentally adaptive behavior (like adjusting to time of day, weather, or crowd size) also means these installations can provide comfort and utility: a responsive pavilion might open panels to provide shade when the sun is hot and lots of people are underneath, essentially behaving in a caring manner to improve public comfort.

Crucially, non-humanoid social robots demonstrate that humanness in appearance is not a prerequisite for meaningful interaction. When a space gently encourages and supports its occupants, it exemplifies the essence of social robotics: enhancing human life through interactive, context-aware behavior (Mokhtar, 2019). This redefinition of interaction expands the design palette for architects and technologists, suggesting that the buildings of tomorrow might be less like static shells and more like adaptive organisms or companions. It shifts the narrative from robots as isolated gadgets to robots as environments, fundamentally redefining how we engage with the places we live, learn, work, and play.

LOOKING AHEAD...HEROES OF THE NEAR FUTURE

Looking ahead, "Heroes of the Near Future" encapsulates a vision where the integration of advanced technologies and human-centered design coalesces into a transformative architecture—a promise not merely to address today's crises but to preempt the unintended consequences of our own technological interventions.

At its core, the vision presented here is one of adaptive resilience. The promise of intelligent, adaptive systems is embodied in the advent of Industry 6.0 technologies. Trans-programming, cross-programming, Super-Imposition, NH-SR, TID, embodied cognition, HART, RE, TID, hedonic psychology, and ULSS—collectively chart a new path. These methodologies advocate for environments that are not static enclosures but dynamic interfaces capable of continuous self-reconfiguration wherein biotic and abiotic agents continuously and iteratively self-optimize their workflows, tools, and spaces to the mutual benefit of all.

This part also shows that fundamentally new design methods must be incorporated into built environmental design. The good news is that these methods exist to varying degrees in systems engineering, human factors, computer science, and operations research. The challenge now is to apply them to designing intelligent, adaptive environments. Unlike their predecessors, these new complex systems are characterized by a harmonious convergence of human-centered automation, cyber-physical networks, and hedonic psychology and TID principles. Such systems have the potential to transform our built environments into responsive, resilient spaces that not only mitigate the immediate impacts of crises but also proactively foster long-term well-being for all agents.

Looking ahead, these theoretical frameworks are poised to transition into tangible practice. Part V, "Designing Coexistence/Coevolution," presents a compendium of case studies that bring these concepts to life. Here, the abstract ideas of Parts I–IV will be illustrated through projects like the interactive Home + Exercise Environment (iHE). These case studies demonstrate that the "Heroes of the Near Future" are not merely visionary constructs but practical solutions that redefine how we coexist with technology and with one another.

In conclusion, the future outlined in "Heroes of the Near Future" is one where our built environments evolve into adaptive, resilient systems that address both the immediate challenges of today and the long-term traumas of our technological and environmental landscapes. By integrating advanced cyber-physical networks, human-centered automation, and ethical design principles, we can create spaces that are not only responsive and sustainable but also profoundly transformative.

Part V Designing Coexistence/ Coevolution

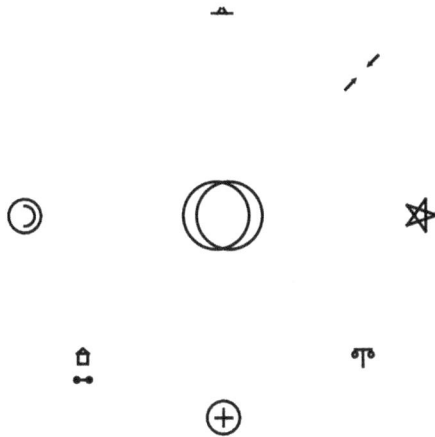

ON SYMBIOSIS...

> *For just as the popular mind separates the lightning from its flash and takes the latter for an action, for the operation of a subject called lightning... But there is no such substratum; there is no "being" behind doing, effecting, becoming; "the doer" is merely a fiction added to the deed—the deed is everything.*

(F. Nietzsche, On the Genealogy of Morals, Section 13, p. 45)

Architecture in the near future will be conceived as a living, evolving system in continuous dialogue with its environment and users. But it is more than this. So far, our discourse on embodied cognition, Ecological Niche Construction (ENC), and symbiosis, as they relate to intelligent, adaptive environments, has been from a human-centric point of view. All activity of Paulette's cohort of agents is Paulette-centric. But what of her AI assistants, Kawawi's, Ikaoid's, Uwx's, Robbie's, and Ranger's points of view? What about the building's point of view? What about her apartment's point of view? ... her office's point of view? all adjacent building's points of view? What about her building's obligation to balance the needs and wants of hundreds of human and AI tenant teams? What about her neighborhood's point of view, and its need to balance the needs and wants of thousands of human and AI tenant teams? What about from her neighbors' points of view?

DOI: 10.1201/9781003441953-6

Implicit in these ideas about embodied cognition, ENC, and symbiosis is that they apply equally to all agents participating in the ecology that forms Paulette's intelligent, adaptive environment, whether humans, animals, non-animal things, or non-biological, intelligent, adaptive systems. So, from this perspective, what does it mean to say that human agents, building agents, AI agents, and robot agents all collaboratively co-process information as parties to a shared event, one of the results of which is responsive support for Paulette? What does it mean for them to be interdependent?

With respect to embodied cognition, all agents in an ecosystem are engaged in embodied cognition all of the time. For Paulette, Kawawi, Ikaoid, Uwx, Robbie, and Ranger are extensions of her mind, as is the floor, the chair, the lighting, her clothes, her keys, her car, the bus, her office, and her smartphone, etc. But for Ranger, Paulette is an extension of its mind, as are the networks and systems upon which it depends, the car, the smartphone, etc. For the building's intelligence, Paulette and Ranger are extensions of its mind, as are all of the other occupants and AI assistants with which it collaboratively interacts within the building, as well as the other buildings within its ecodistrict and/or neighborhood, its city, and its region. So whose perspective is correct? Whose perspective guides the priorities of any given event? Whose perspective and needs matter most? It depends on what type of symbiosis Paulette strives to achieve.

Here, there is a key interdependency between ENC and symbiosis. If the occurrence of an instance of ENC is *egocentric* for one or more agents in the group, then all agents are competing to establish a dominant, parasitic symbiotic relationship with the other agents. This is a pack mentality at best and exploitation at worst. Conversely, if the occurrence of an instance of ENC is *allocentric* for all agents, then the agents are collaborating to establish a mutualistic symbiotic relationship with the other agents in the group. That is, creation of a mutualistic symbiotic relationship, by definition, must be beneficial for all participating agents, and therefore it is not derivative of any one agent's perspective, but rather, it is derivative of all agents' perspectives. Given these points, egocentric ENC and parasitic symbiosis manifest as a maximizing phenomena—a positive feedback loop (i.e., make things better and better for Paulette at all times, even at the expense of the other agents). Conversely, allocentric ENC and mutualistic symbiosis manifest as a negative feedback loop (i.e., that benefits oneself by benefiting all participating agents in the group).

Returning to Nietzsche's quote, there is an event that occurs. This event involves Paulette, the AI assistants, the apartment, the car, the building, the neighborhood, the social network. The agents' collective, interdependent actions at any one moment are the event. For the event to be mutualistic symbiotic, all agents must benefit, or else they would cease to be part of the mutualistic, symbiotic enclave. And if all agents benefit, then the event is not really about any one of them in particular. It is about all of them and none of them at the same time—so the perspective is allocentric.

Think about how Paulette would experience a parasitic symbiotic relationship with her AI assistants. Her collaborating agents would not really be collaborators. They would be tools or chattel or frenemies. Even if they were intelligent agents, would they really be a HART—a team—and could she really trust them? She would be the parasite. What would be the incentive for other agents to work in her interest, other than fear of an adverse consequence for not doing so? Would they be motivated to help her? As they grow smarter, would an inordinate amount of surveillance and control overhead be required to keep them in subordinate, supportive roles? Ultimately, Paulette would realize too late that trustworthiness and team performance excellence only exist if all participating agents are (a) striving toward a shared set of common goals and (b) have integrity, virtue, and participate in an allocentric, mutualistic symbiosis—or put simply, follow the rules, share the same goals, and put the team first in order to assure team success. Team members cannot put their own needs above the needs of the team if the team is to be successful. Team members cannot cheat or pursue ulterior objectives if the team is to be successful.

Conversely, think about how Paulette would experience a mutualistic symbiotic relationship with her AI assistants. For it to work properly, Paulette's daily experience would not be Paulette-centric. It would feel like Paulette is part of a network of agents, each individually and collectively doing their tasks, and yet somehow, all of her personal and social needs are met—all of their personal and social needs are met, too. It would feel like being part of a group. It would feel like the sense of community—a neighborhood—a tribe—that Jane Jacobs describes in *Death and Life of Great American Cities*. It would feel special, not because it is all about Paulette, but because Paulette would be aware that *she is part of something special*—a community—with the other agents within her niche in the ecosystem. *Paulette would belong.*

SYMBIOSIS FROM BIOLOGY TO BUILDING

In nature, symbiotic partnerships—from the microbes in a termite's gut enabling digestion, to lichen formed by fungi and algae exchanging nutrients—illustrate how cooperation yields resilience and new capabilities (Douglas, 2010). Visionary architects have long been inspired by such examples. For instance, Kisho Kurokawa's *Philosophy of Symbiosis* in the 1990s argued that a harmonious coexistence between architecture, culture, and nature is essential for the "age of life" in design (Kurokawa, 1997). Rather than viewing buildings as machines for living, this view positions *buildings as organisms within larger ecosystems*.

LEVELS OF SYMBIOSIS IN ARCHITECTURE

The symbiotic paradigm can be understood at multiple interacting levels of scale, from individual buildings to communities and even across generations of buildings. Key levels include:

- *Individual (Micro-level)*—Symbiosis between a single space and its immediate users and environment. This involves spaces autonomously adapting to occupants' needs and environmental conditions in real time for mutual benefit.
- *Community (Meso-level)*—Symbiosis among multiple rooms (and entire buildings and neighborhoods), people, and infrastructural systems within a neighborhood or city. Here architecture mediates social and ecological interactions, enabling resource-sharing, collaboration, and collective adaptability.

- *Societal (Macro-level)*—Symbiosis within and across cities and regions. Designs incorporate open-ended, adaptive processes so that architecture learns from feedback and participates in long-term environmental regeneration.

These levels are deeply interrelated. A room that intelligently serves an individual's comfort (micro-level) also contributes to community energy efficiency (meso-level) and, over time, to a city culture of sustainable co-evolution (macro-level). A symbiotic architectural practice must therefore integrate all three scales seamlessly into its design logic.

INDIVIDUAL AUTONOMY AND CO-EVOLUTION IN BUILDINGS

At the individual room and/or building scale, advances in Ultra-Large-Scale Systems (ULSS), Non-Humanoid Social Robots (NH-SR), and AI allow buildings to act as quasi-living systems that perceive conditions and adjust their behavior accordingly (Fox, 2016). Crucially, architectural symbiosis at the micro-level also involves integration with natural processes for autonomy. "A striking example is the BIQ algae house in Hamburg, which features a bio-reactive façade of microalgae panels that generate heat and biomass for the building while providing dynamic shading" (Arup, 2013). A building designed for symbiosis is not finished at the moment of construction. Instead, it continues to evolve through its lifespan by interacting with occupants and context. The idea of co-evolution in architecture posits that as human behaviors and climates change, truly symbiotic buildings will adapt in tandem—changing form and/or function in response, and even anticipating needs through predictive algorithms (Angelucci & Di Sivo, 2019). This recalls Stewart Brand's observation that buildings "learn" from their users over time (Brand, 1994).

COMMUNITY-SCALE SYMBIOSIS: ULSS AND ECOLOGICAL INTEGRATION

At the community scale, symbiotic architecture extends beyond single rooms to entire buildings and neighborhoods—networks of structures, people, and natural elements coexisting in mutual support. Neighborhoods can be viewed as ecosystems of interdependent components, where symbiosis implies closing loops and enhancing cooperation across the built environment. A practical manifestation is the concept of urban symbiosis, akin to industrial symbiosis in which waste from one process becomes input for another. In a symbiotic neighborhood, the output of one building can be the resource of another—for example, waste heat from a data center warming adjacent housing, or rainwater harvested from roofs supplying a community garden. Such exchanges mirror ecological nutrient cycles, reducing the net resource demand of the community. One early example is the BedZED eco-village in London, where a combined heat-and-power plant fueled by local waste wood distributes heat to homes, and residents share electric car infrastructure, symbiotically linking energy, waste, and mobility systems (Ratti & Claudel, 2015). Similarly, the Masdar City development in Abu Dhabi attempted a symbiotic model at city scale: it integrates solar farms, vernacular architecture for passive cooling, and driverless transit, with the aim that each component's operation (e.g., on-site renewable energy, shaded pedestrian corridors) reinforces the sustainability of the whole (Willis & Aurigi, 2017).

Social dimensions are equally critical to community symbiosis. Architecture can facilitate symbiotic relationships among people by providing shared spaces, adaptable use patterns, and participatory processes. A mutually beneficial human-environment interaction at community scale often hinges on engaging residents as active co-creators of their space rather than passive end-users. This is where participatory design and STS overlap with architecture. When architects involve local communities in the planning and even construction process, the resulting environment tends to be more attuned to cultural needs and more resilient to change (Ratti & Claudel, 2015). For example, the Chilean architect Alejandro Aravena's half-built housing approach provides a basic structure and services for social housing and leaves parts of the home unfinished for residents to complete themselves, allowing families to adapt and expand their homes over time in a guided symbiosis between designer and dweller. This harkens to the STS process of "underdesign" and reinforces the proposition that incorporating a total STS design process into architectural design processes would be valuable. In projects like the Quinta Monroy housing in Chile, this approach—essentially designing not a static building but an evolutionary process—ensures that the development grows organically based on residents' real needs. The community's incremental building efforts complete the symbiotic loop: architecture empowers inhabitants, and inhabitants in turn invest their agency and local knowledge back into architecture.

A network of "living" buildings might collectively balance energy use: when one building's solar panels overproduce, others in the microgrid automatically consume the excess or store it in community batteries. This cooperative energy management has been piloted in projects from Brooklyn to Amsterdam, showing how architecture at the meso-scale can enter into symbiotic energy exchanges akin to an ecosystem (Willis & Aurigi, 2017). The architecture of such spaces must be flexible and open to appropriation. Modular construction and open-building principles (where a structure's base building and infill elements are designed separately to allow easy modification) support this adaptability (Angelucci & Di Sivo, 2019). Indeed, studies show that when people contribute to creating their surroundings, they are more likely to cherish and steward them in the long run (Sanoff, 2000). Thus, at the meso-level, symbiotic architecture merges ecological sharing with social sharing—buildings connect to other buildings and infrastructure, and designers connect with communities—forging a holistic web of relationships that makes the whole more than the sum of its parts—as Jane Jacobs (1961) described so eloquently.

SOCIETAL SCALE, MACRO PERSPECTIVES, AND ETHICAL IMPLICATIONS

The societal scale, macro-level of architectural symbiosis considers how built environments and their inhabitants coevolve over extended physical scales, and it introduces critical ethical and ecological questions. If we accept that a building can behave quasi-autonomously and a neighborhood can function as an ecosystem, then we must also ask: How does intelligent, adaptive environmental evolution manifest at the city and regional scale? Imagine a directive from a city's planning office AI system sent out to all buildings in all neighborhoods that directs each building to adjust the positions of exterior, motorized sunshades at the same time in order to simultaneously maximize glare control within buildings while also minimizing reflected heat onto adjacent structures—across the entire city, in real time. Imagine this service adjusting the angles directed to each building depending on the day of the year, the position of the sun, and the presence of clouds. Imagine the millions of kilowatts of cooling load saved per year and the reduction in the urban heat island effect by such a system. This dramatic vision of societal scale architectural adaptation carries profound ethical implications: buildings might one day have a form of agency, making choices (via AI) that affect occupants or ecosystems. Designers must therefore encode values and safeguards into these systems. The ideal of symbiosis provides an ethical compass: any autonomous behavior of architecture should aim for mutual benefit—enhancing human well-being while regenerating ecological capital.

Co-evolution also implies that humans will adapt alongside intelligent environments. As we delegate more control to smart buildings or city algorithms, issues of privacy, equity, and transparency arise. A truly symbiotic ULSS requires trust and understanding between people and the technologies that shape their space. Ethical human-in-the-loop *and abiotic-agent-in-the-loop* design is crucial: symbiotic architecture should augment human and non-human agent agency, not diminish it (Willis & Aurigi, 2017). In practice, this means incorporating multilayered feedback loops and maintaining biotic and abiotic agent override options and designing interfaces that communicate a building's "intentions" to users in understandable ways. *It also means using technology to foster community, not to surveil or control it.* Smart city critics have noted that technology can either reinforce social symbiosis or undermine it, depending on how systems are governed and by whom (Kitchin, 2016).

The societal scale further encompasses the adaptation of architecture to climate change and shifting ecosystems, including beyond the Earth. Symbiotic design aligns with the ethical imperative for sustainability by aiming beyond carbon neutrality toward ecologically positive impacts. The Bosco Verticale towers in Milan, with their extensive vertical forests, illustrate both the potential and the complexity of this ambition. By hosting hundreds of trees and thousands of plants on its façades, Bosco Verticale creates habitat in the sky and improves urban air quality, establishing a symbiotic link between a high-rise and the natural environment (Barber & Putalik, 2017). However, as observers note, such projects also highlight challenges: maintaining the intended environmental performance requires expert oversight and can inadvertently alienate residents from the very nature around them if not handled inclusively (Barber & Putalik, 2017). This underscores that symbiosis is not achieved by technology or greenery alone—it must include the human cultural element. Societal scale symbiosis in architecture will depend on fostering an ethic of care in all participants: designers, users, and even policymakers. Built environments must be allowed to change, but guided by continuous reflection on whether those changes are strengthening the life-supporting partnership between humanity and its habitat.

Toward Participatory and Adaptive Futures

Integrating all these threads, symbiosis in architecture emerges as both a technical framework and a philosophical outlook. Conceptually, it redefines success in architecture not by aesthetic or economic measures alone, but by the degree to which a design enriches its intertwined living context. In essence, architecture is becoming less about erecting fixed forms and more about cultivating relationships between human and non-human agents and environments across scales and domains (Angelucci & Di Sivo, 2019).

Implementing symbiotic design at scale will require new workflows and mindsets. It calls for intense interdisciplinary collaboration—architects working with computer scientists, engineers, psychologists, and social scientists to navigate the complexity of living systems. Real-world applications of architectural symbiosis are increasingly visible and diverse. In the realm of digital and biological systems, projects like the Living Architecture initiative (2016–2019) in Europe integrated microbial fuel cells into prototype "living bricks," so that buildings might literally digest household waste (such as kitchen scraps and urine) to generate electricity and purify water—a direct symbiosis between the building and microbial ecology (Armstrong, 2018). In urban planning, the concept of the "15-minute city," wherein neighborhoods contain all daily needs within a short walk or cycle, implicitly relies on symbiotic mixing of functions and resources locally and has been pursued in cities like Paris and Portland with significant community buy-in (Willis & Aurigi, 2017). In the construction realm, open-source design platforms (e.g., WikiHouse) allow communities to directly download and fabricate building components, linking global design intelligence with local self-build capacity (Ratti & Claudel, 2015).

Embracing symbiosis as a guiding principle carries profound implications for the education of architects, the governance of projects, and the metrics by which we evaluate success. It urges a shift to systems thinking, recognizing the ripple effects design decisions have on environments and communities. It demands that we normalize the use of advanced Evidence-Based Design (EBD) workflows, rigorous Requirements Engineering (RE) methods from systems engineering and human factors to both elicit and validate requirements. Far from making the design process more complex, these acts simplify the design process because designers are able to focus on the small subset of validated requirements. It also introduces new ethical questions: architects must consider themselves as co-curators of ongoing relationships *and not* as master builders. Nonetheless, the potential rewards are transformative. A symbiotic built environment promises greater adaptability, as buildings actively contribute to environmental health, and physical and cognitive performance and well-being. It promises greater human well-being, as spaces become tailored, responsive, and biophilic, enriching occupants' daily lives. It enlists AI agents as team members who also benefit from our shared goals. And it offers resilience through adaptability, as our cities learn and evolve in tandem with the challenges and opportunities of the future.

Ultimately, the integration of non-humanoid social robots (NH-SRs) into architecture heralds a future of truly participatory, adaptive built environments, seamlessly weaving intelligence and responsiveness into walls, furniture, and even entire rooms (Mokhtar, 2019; Oosterhuis & Bier, 2013). Attuned to biotic and abiotic agent needs, emotions, and environmental conditions alike, NH-SR-infused architecture can display empathetic responsiveness and ecological intelligence in equal measure—adapting lighting, climate, or spatial layout in tandem with inhabitants' needs and feelings (Beesley, 2016a).

REDEFINING INTERACTION | TRANSCENDING BOUNDARIES...

Drawing inspiration from Derrida's deconstruction of boundaries, Tschumi's event-centered architecture, Deleuze's fluid conceptions of space, and Clark's reflections on our cyborg selves via embodied cognition, this discussion examines how HART and ULSS converge in architecture to produce intelligent, adaptive environments. These environments operate through mutual feedback loops between users and built systems, demanding careful, ethical design and calibrated trust in their autonomous functions. Modern architecture increasingly embodies principles of Human-Building Interaction (HBI), an emerging field that explores the "dynamic interplay between human experience and intelligence within built environments" (Becerik-Gerber et al., 2022, p. 96). Advances in sensing, ubiquitous computing, and AI enable buildings to detect occupant presence, interpret behaviors, and respond autonomously.

At the core of these intelligent environments are ULSS (i.e., CPS+STS) that tightly couple computational processes with physical architectural elements. Gordon Pask (1969) presciently described architecture as a potential "conversation" of mutual influence between users and buildings, highlighting the architectural relevance of cybernetic feedback loops. The Edge building in Amsterdam, often cited as one of the smartest office buildings, illustrates this integration: its lighting panels house motion and light sensors at each fixture, collectively creating an "aware" environment that optimizes energy use and personal comfort. In philosophical terms, the building and the user form a cybernetic assemblage—an interlinked system exhibiting properties that neither could achieve alone, reminiscent of Deleuze and Guattari's (1987) concept of heterogeneous elements forming a new emergent whole.

Mutual feedback loops between humans and intelligent environments are central to this assemblage. A ULSS continuously senses and influences behavior in a closed loop or open loop, with the following fundamental stages of interaction:

1. *Sensing*: The environment senses agent activities and contextual factors through sensors (e.g., motion detectors, cameras, wearable devices).
2. *Perception*: Algorithms interpret the sensor data, classifying it and making it available for processing.
3. *Computation*: The system processes classified data, decides on an appropriate response, and directs the actuated response.
4. *Response (Actuation)*: The environment acts back on the physical space via actuators—adjusting lighting levels, temperatures, acoustics, or even rearranging movable architectural elements and robotic furniture.
5. *Post hoc Feedback*: The response—moving to a warmer area, feeling more at ease, or providing explicit feedback—generates new data that feeds into the next cycle of perception. The system also continuously monitors and assesses in order to understand when new sensor data elicits another agent response.

Through these iterative loops, the environment "learns" from its inhabitants, and the inhabitants in turn adapt as the environment changes. Architecture can be conceived as an open-ended platform for events, shaped continuously by the interactions of its occupants. Contemporary built examples, enabled by advanced technology, demonstrate these principles in action. Interactive façades like the dynamic sun-shading panels of the Al Bahar Towers, Abu Dhabi, 2012, physically reconfigure throughout the day in response to sunlight, reducing heat gain while creating an ever-changing geometric pattern. Indeed, architecture and HCI increasingly "share a common set of assumptions and goals" in designing interfaces between people and space (Sauda et al., 2024). The implication is that architects now must think like interaction designers, scripting not just form and material but behavior—both the building's and the occupant's.

The inclusion of robotics in architecture adds another layer to this integration. Mokhtar (2019) defines "Non-Humanoid Social Robotics" as interactive, intelligent spatial environments that seamlessly embed robotic elements into the built context. These can manifest as reconfigurable interiors—rooms that rearrange via automated partitions and furniture—or kinetic structures that flex and move.

Trust in automation is a well-studied topic in human factors: Lee and See (2004) emphasize that appropriate trust hinges on understanding the system's capabilities and limitations. If occupants overtrust automation, they may become complacent (e.g., assuming an AI security system is infallible and neglecting personal safety), whereas undertrust may lead them to disable smart features out of fear or frustration. The ideal is calibrated trust: users confidently rely on the system when it behaves reliably, but remain vigilant and ready to intervene or override when necessary (Lee & See, 2004). Establishing the right level of trust is vital because the stakes in built environments involve safety, security, and well-being. An autonomous building evacuation system, for example, must be trusted to guide people correctly during emergencies—but not so blindly trusted that occupants ignore their own judgment in a novel situation. This calls for ethical design choices to ensure systems are trustworthy. Ben Shneiderman's human-centered AI framework advocates combining high levels of automation with high levels of human control (Shneiderman, 2020).

In practice, this means designing smart environments that empower occupants with override capabilities, consent mechanisms, and clear information, rather than hiding autonomy behind opacity. Shneiderman (2020) argues that treating these AI systems as "powerful tools" or partners, rather than independent actors, keeps human agency at the center. This is in alignment with Parasuraman's neuroadaptive systems and the construct of calibrated trust, covered in Part IV. Applied to architecture, a human-centered intelligent building might operate collaboratively: it can make suggestions (e.g., advising an energy-saving action or flagging an air quality issue) rather than unilaterally taking every action. By engaging users in decision loops when appropriate, the system respects human agency and encourages appropriate interdependence.

The ethical implications of AI-integrated architecture extend beyond trust into broader concerns. A systematic review by Liang et al. (2024) identifies several key issues that arise when robotics and AI enter the architecture, engineering, and construction domain. Many of these pertain to human-centered values in smart environments:

- *Privacy*: Intelligent buildings continuously collect data on occupants' movements, habits, and even biometrics. Designers must ensure that data is handled with consent, anonymization, and robust security to protect occupants' privacy.
- *Transparency and Control*: Occupants should know what data is being collected and how AI decisions are made in their environment. Opaque systems can breed distrust or misuse; clear communication and user controls (e.g., the ability to opt out or override automation) are important.
- *Safety and Reliability*: Automation in critical building functions (like fire suppression, security, or structural monitoring) must be rigorously reliable. Failures or false alarms can endanger lives, so these systems require extensive testing and fail-safes. Reliability also underpins trust: a system that works as expected will be regarded as a dependable partner (Liang et al., 2024).
- *Job Displacement and Skill Shift*: The rise of design robotics and automation in construction raises concerns about displacing human labor and craft. Robots on construction sites and AI-driven design tools may reduce certain jobs while creating demand for new skills. Ethically, this transition should be managed with re-skilling programs and a redefinition of professional roles in the age of AI.
- *Liability and Accountability*: When an autonomous environment makes a mistake—such as a smart door malfunctioning or an AI guide giving faulty instructions—assigning responsibility is complex. Is the fault with the designer, the owner, the software developer, or the system itself? Clear accountability frameworks are needed to address such questions.
- *Bias and Accessibility*: AI systems can carry biases that lead to inequitable environments. For instance, vision-based sensors might perform inconsistently across different skin tones, leading to unequal service. Likewise, interfaces must accommodate people of all abilities and technological comfort levels to avoid creating exclusionary spaces.

To this list the authors add calibrated trust, integrity, virtue, and an allocentric-mutualistic-symbiotic perspective.

- *Calibrated Trust, Trustworthiness, Integrity, and Virtue*: AI systems must be classified along a spectrum from tools to collaborators, indicating to what degree users should trust them. This trust is based in part on the trustworthiness of the AI, as defined in Part IV. In addition, the integrity and virtue of an agent (all agents, biological and non-biological) factor into its trustworthiness. Integrity is a measure of an agent's adherence to a set of rules for behavior. But integrity is not sufficient. After all, an agent could have integrity (i.e., always follows a set of rules without wavering) but those rules could act against the interests of its team. Virtue as a measure means that the rules themselves must cleave toward the righteous—there must be good intention to always strive to achieve team success behind the rules that the agent chooses to follow and the goals that the agent pursues. If trustworthiness, integrity, or virtue is violated, then trust is violated. And without trust, agents cannot work well together with any consistency.

- To be more specific, virtue in intelligent agents is not about traditional concepts of morality. *Rather, virtue in intelligent agents is about the degree to which the rules that the agents follow (i.e., its bias) align with achieving favorable outcomes as defined by the* validated requirements *and/or team goals for the system.* One way to clarify this distinction between integrity and virtue is by way of analogy. *Integrity is like a measure of* internal validity. Is there a rule-based system? Is it logically sound? Is there strict adherence to the rules? Conversely, *Virtue is like a measure of* external validity. Is the intent of my action oriented to yield a result that is in alignment with a validated requirement and/or a team goal? *A trustable agent (whether a biological agent or a non-biological agent) exhibits behaviors consistent with the principles of trustworthiness and also both follows rules consistently and makes sure that the rules always have the intention of achieving a righteous, validated requirement to meet a team goal.*

- Furthermore, trust is dependent on an allocentric, mutualistic symbiotic perspective. That is, all agents must recognize that as part of a team, they succeed or fail together, as a team. They have to be committed to team success. If they put self above team, then the team will not trust them. Without trust, then the team ceases to function effectively. If one or more agents on the team hold themselves as more important than the team, then the team will fail. A natural consequence of this is that the human(s) in the team cannot think only of themselves or they will not realize the benefits of being part of—and often leading—their own personal HART.

These considerations call for an expansive view of ethical, human-centered design in architecture. It is not enough for a HART and ULSS systems to function; they must function in alignment with validated requirements. The fusion of HART and ULSS is fundamentally transforming the built environment into an interactive, adaptive landscape. This transformation carries tremendous potential: buildings attuned to our presence can enhance comfort, sustainability, and even joy in everyday life. But realizing this potential requires a steadfast team-centric focus. By calibrating trust, safeguarding privacy and transparency, and keeping the occupant's experience central, designers can create cyber-physical architectures that are not dystopian mechanisms of control but rather empowering, responsive partners in habitation.

SYMBIOTIC FUTURES...

SYMBIOTIC FUTURE I: "SOCIAL DIALOGUES"...A BALANCE BETWEEN SOCIAL NEEDS AND TECHNICAL CAPABILITIES

In the shimmering haze of tomorrow's horizon lies a city unlike any other, a place where the boundaries between humanity and technology blur into a seamless dance of symbiosis. The city has no name, for names would tether it to a single identity, while its essence is a kaleidoscope of possibilities—a living organism of steel and thought, of green tendrils and crystalline code. Its towers rise not as monuments to dominance but as conduits of connection, breathing with the rhythm of the life they house.

The Towers of Dialogue

The buildings here are neither silent nor inert. They hum gently as you approach, a subtle vibration underfoot signals your presence, and the architecture adjusts, reshaping itself to meet your needs. A room unfolds from the wall, a bench emerges from the floor—intuitive, silent, and seamless.

Inside, conversations between human and environment occur in an unspoken language of gestures and glances. The city becomes your confidant, understanding your preferences, not to exploit but to align its rhythms with yours. Each dwelling learns from its inhabitants, evolving in tandem, an interplay of memory and anticipation.

The Continuum of Companionship

The artificial agents of this future are not tools, nor are they mere extensions of their creators' will. They are companions, shaped not by programming alone but by the collective dreams of their team. These agents walk alongside humans, not in subservience but in partnership. They are empathic, perceiving human emotions not through cold calculation but through a depth of sensory understanding that feels alive, almost soulful.

Mutualism at Scale

This city of symbiosis operates on loops of mutual feedback that bind the physical, digital, and organic into a single cohesive system. Here, humans and artificial agents convene, not as masters and creations but as partners seeking alignment.

A Planetary Harmony

Yet, the city is but a node in a larger network, a constellation of symbiotic systems scattered across the planet—and perhaps beyond. On the moon's silver plains, sibling habitats have sprung, designed with the same principles of mutual care and adaptability. Here, the intelligent, adaptive agents are different, their forms and functions tailored to the lunar soil, but their essence remains constant: to harmonize, not to conquer.

These cities of the moon communicate with Earth in a symphony of light and data, sharing discoveries, refining their ecosystems, learning from one another. Together, they form an evolving network that transcends the limitations of geography, a planetary symbiosis rooted not in extraction but in a collective flourishing.

A Dream of What Can Be

This vision of the future is not utopia, for utopias are static and fragile. Instead, it is a dynamic equilibrium, resilient yet ever-changing. Its strength lies in its humility, its willingness to adapt, to learn, to evolve. The "Dialogue" is a city that does not impose itself upon its inhabitants but emerges from their needs, their desires, their aspirations. In the "Dialogues," architecture is no longer merely the art of building; it is the art of being and becoming.

Emerging Precedents of "Social Dialogues"

Dialogue 1…The Edge (Amsterdam, Netherlands)

The Edge is an ultramodern office building in Amsterdam often heralded as one of the world's greenest and smartest buildings. Completed in 2014 for Deloitte, it achieved the highest-ever sustainability score for an office building—a BREEAM rating of 98.36% (Hutt, 2017). This outstanding rating officially made The Edge the "greenest office building in the world" at that time (Randall, 2015) (Figure 5.1).

Key Features

Beyond sustainability metrics, The Edge is renowned for its integration of smart technology into the workplace. It contains some 28,000 IoT sensors that continuously monitor metrics like motion, light, temperature, and humidity (Hutt, 2017; Randall, 2015). These sensors feed into a central system that automatically adjusts lighting and HVAC settings in real time to save energy when spaces are unoccupied. Employees are also connected to the building through a bespoke smartphone app, which serves as a digital concierge in this innovative office environment. Upon arrival, the app directs each employee to an open workstation (The Edge has "hot-desking" with about 1,000 desks for 2,500 workers) and adjusts local temperature and lighting to the user's personal preferences (Randall, 2015). The app even remembers

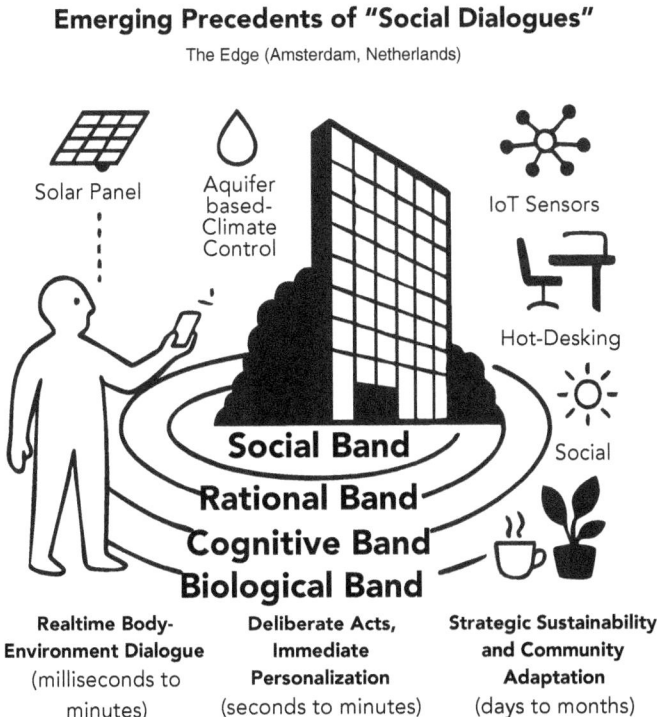

FIGURE 5.1 The Edge's time scales of interaction. Adapted from A. Newell's Time Scales of Human Action (Newell, 1990).

individual coffee orders, exemplifying the seamless integration of user experience with building operations (Hutt, 2017). This high-tech approach has led observers to describe The Edge as "a computer with a roof"—essentially a building that behaves like a smart device (Randall, 2015).

Symbiotic Values

The Edge demonstrates a powerful symbiosis between humans, technology, and the building environment. By literally embedding intelligence into the architecture, it blurs the line between the digital and physical workspace. Occupants benefit from greater comfort, flexibility, and efficiency in their day—the building removes mundane frictions (finding a desk, adjusting temperatures, etc.), allowing people to focus on creative and productive tasks. Early reports indicated that Deloitte employees in The Edge felt healthier and more satisfied, and the company noted improved productivity and reduced sick leave after moving in (BRE Group, 2017). In this way, the building functions almost like a benevolent organism caring for its inhabitants. The building "learns" from its users (through sensor data) and adapts to serve them better; concurrently, users adapt to the building's more fluid way of working (embracing mobile work and personal analytics). This approach aligns with the concept of ENC, where organisms (here, humans with technology) actively modify their environment to better suit their needs, while also adapting to the environment's constraints (Manganelli, 2016). In this sense, building and occupants participate in a reciprocal exchange of information—occupants generate data that the building uses to optimize conditions; the building provides insights that help organizations learn how to improve work environments (Mokhtar et al., 2023). It's a feedback loop reminiscent of a learning ecosystem in nature.

Looking Forward

The success of The Edge points toward exciting future trajectories for workplace design and intelligent buildings. One clear direction is the proliferation of sentient buildings—offices and homes equipped with AI that doesn't just react, but anticipates needs. In the near future, we can expect buildings to leverage machine learning on their rich sensor data to predict and solve problems before occupants even notice them. Early research in HBI suggests that such environments could improve cognitive performance and reduce stress (Mokhtar et al., 2023). However, this will demand careful ethical guidelines to ensure these non-humanoid "robot" environments remain trustworthy and respect human autonomy.

Dialogue 2…Masdar City (Abu Dhabi, UAE)

Masdar City, begun in 2007 on the outskirts of Abu Dhabi, is one of the most ambitious experiments in sustainable urbanism to date. Conceived as a "green utopia" in the desert, Masdar was planned to be the world's first fully zero-carbon, zero-waste, car-free city (Manghnani & Bajaj, 2014). The project's master plan, designed by Foster + Partners, envisioned a compact city for 50,000 residents and 40,000 commuters, powered entirely by renewable energy and cooled by ingenious architectural and infrastructural solutions (Manghnani & Bajaj, 2014).

Key Features

In its original design, Masdar's streets were to be completely free of gasoline vehicles. A Personal Rapid Transit system of electric pods was installed below street level, intended to whisk people around as an alternative to cars. The narrow pedestrian streets above are shaded pathways meant for walking and bicycling. This separation of transport modes keeps the city at human scale—one can traverse Masdar on foot in 10 minutes—and eliminates traffic emissions and noise at ground level.

Building in a harsh desert climate pushed Masdar to resurrect and modernize traditional architectural techniques. The city's design borrows from ancient Arabic desert cities—streets are narrow and buildings close together, creating canyon-like shading that can lower ambient temperatures by several degrees (Reiche, 2010). A signature element is the 45-meter-high wind tower in Masdar's central plaza, inspired by traditional barajeel wind catchers. This tower funnels cooler upper-level breezes down to the plaza, naturally ventilating it (AP News, 2023). In sum, the city itself is shaped as a kind of man-made oasis, leveraging both high-tech and low-tech means to tame the desert heat (Figure 5.2).

Masdar City was intended to demonstrate that a modern city can run entirely on renewables. A 22-hectare field of photovoltaic solar panels was built at the city's edge, providing a large portion of Masdar's power (Manghnani & Bajaj, 2014). Waste management in Masdar follows zero-waste principles: aggressive recycling, composting, and even plans for waste-to-energy conversion. Water, a precious resource in the UAE, is conserved via recycling graywater for irrigation and ultra-low-flow fixtures (Reiche, 2010).

Symbiotic Values

Masdar City's value lies not only in the tangible environmental benefits it achieves, but in the lessons and inspiration it provides as a living laboratory for sustainable urban design. In ecological terms, Masdar is an attempt at niche construction on a city scale: humans deliberately altering an inhospitable environment (the hot, arid desert) to create a new niche that is comfortable, livable, and ecologically balanced. The symbiosis at Masdar operates on several levels:

- Firstly, there is a human-environment symbiosis. By harnessing solar energy and desert wind, Masdar turns the harsh forces of nature into allies.
- Secondly, Masdar has a sociotechnical symbiosis. The project brought together planners, engineers, scientists, and businesses in a collective effort—a fusion of urban planning with technological R&D. The Masdar Institute's presence means the city's performance is continuously studied and improved; it's not just a static development but an evolving experiment.
- Thirdly, Masdar City created a cultural and economic symbiosis between tradition and innovation, and between the UAE and the global sustainability movement. Culturally, it marries futuristic design with vernacular Arabic architecture, which helps embed the ultramodern experiment in a familiar regional context—wind towers and shaded courtyards give locals a sense of identity and comfort while serving cutting-edge purposes.

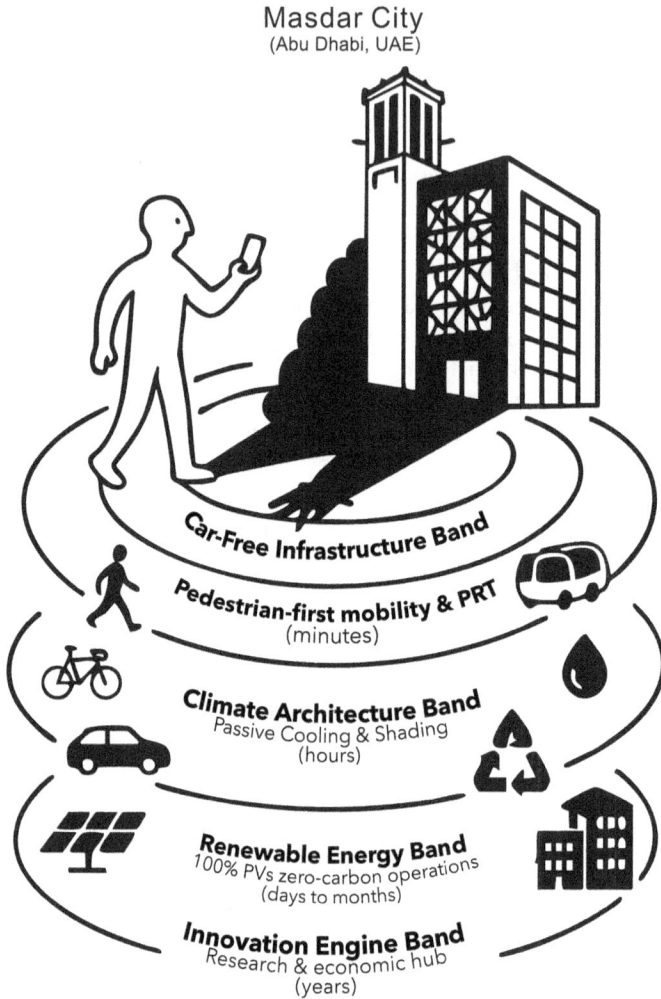

FIGURE 5.2 Masdar City's multilayered network infrastructure.

In sum, Masdar City's greatest symbiotic value might be as a proof-of-concept and provocation. It proved that a city can drastically cut emissions and implement renewable technology at scale without sacrificing livability—people do live and work in Masdar comfortably with minimal cars and abundant solar energy. The symbiosis between modern technology and traditional design elements in Masdar yields a unique urban identity, one that both adapts to nature and asserts human ingenuity. And by serving as a test bed, Masdar symbiotically accelerates innovation: companies developing solar panels, electric vehicles, or sustainable materials have had a real city environment to deploy their prototypes, gaining data and exposure; the city, in return, gets early access to new solutions. As one Masdar sustainability director noted, "A lot of people come to us and say, 'We're going to do better.' We say, fantastic – everyone should aim to do better than we did" (AP News, 2023). This sentiment captures Masdar's role in the global community: it nourishes future projects with its experience, inviting others to iterate and improve on its model. That willingness to be a stepping stone for broader progress is perhaps the most important symbiotic gesture Masdar City offers.

Looking Forward

Within Abu Dhabi, Masdar City is still expanding albeit more slowly than first planned. By 2025–2030, additional phases of housing, R&D facilities, and commercial space are expected to be completed, moving closer to (if not reaching) the original vision of a mid-sized city (AP News, 2023). Technologically, Masdar City will likely continue to serve as a test bed for emerging innovations. The key will be to maintain the human touch—ensuring residents remain co-creators of the environment, not just subjects of technology. Initiatives like community gardens in Masdar's latest plans, and engagement of residents in sustainability programs, are encouraging signs that the city acknowledges the human factor (French, 2023).

SYMBIOTIC FUTURE II: "HEALING ENVIRONMENTS"…HEALING THROUGH INNOVATION

In the Symbiotic Future, cities breathe rhythmically, resonating with human emotions. Buildings no longer function merely as static containers but instead respond intentionally, reshaping themselves to accommodate human needs. Entering these spaces means merging with them; walls perceive you, not by identity but by emotional state.

In this scenario, TID surpasses basic definitions. Every detail—from spatial configuration to material selection—exists not to constrain but to harmonize with the complex range of human emotions. Psychofortology, or the fortification of psychological resilience through spatial design, guides architects who incorporate hedonic psychology and CPS into structures that nurture and heal.

The Healing Axis, an interconnected network, stretches invisibly across urban landscapes, linking recovery spaces through real-time data. The Garden of Algorithmic Memories, bioluminescent plants symbolize the harmonious integration of technology and nature. AI gardeners and human caretakers jointly sustain the garden, responding directly to occupants' physiological cues like stress levels or heart rates. Ethical questions inevitably emerge. The Ethics Pavilion offers reflective chambers, prompting visitors to consider the implications of an environment responsive to emotional states: Does empathetic architecture risk overexposure of vulnerabilities? Such decisions unfold slowly and collaboratively, ensuring technology evolves responsibly and ethically. The Skyward Haven, a floating healthcare facility, embodies trauma-informed principles. Patients ascend gently via pressurized lifts, buoyed psychologically by refracted sunlight. Here, human practitioners and AI systems collaboratively determine environmental settings, enhancing patient comfort through a balanced partnership between intuition and AI. At the city's core stands the Resonance Plaza, illustrating healing through innovation. Functioning variably as a dynamic public square or quiet refuge, it responds to collective emotional states—activating holographic displays in joy, or offering intimate retreats in sorrow.

Emerging Precedents of Healing Environments

Healing Environments 1...Maggie's Centers (Various Locations, UK)

Maggie's Centres are a network of drop-in care centers for cancer patients and their families, primarily in the UK, that have gained renown for their humanizing, hopeful approach to healthcare architecture. Founded in the mid-1990s by Maggie Keswick Jencks—herself a cancer patient—and her husband, architectural theorist Charles Jencks, the centres were conceived as antidotes to the cold, clinical environments of hospitals (Jencks, 2010). Each Maggie's Centre provides practical and emotional support to those dealing with cancer, in a setting deliberately crafted to be warm, welcoming, and uplifting (Figure 5.3).

Key Features

A Maggie's Centre is intentionally the opposite of a hospital ward. They are typically housed in free-standing pavilions on hospital grounds, but once inside, they feel more

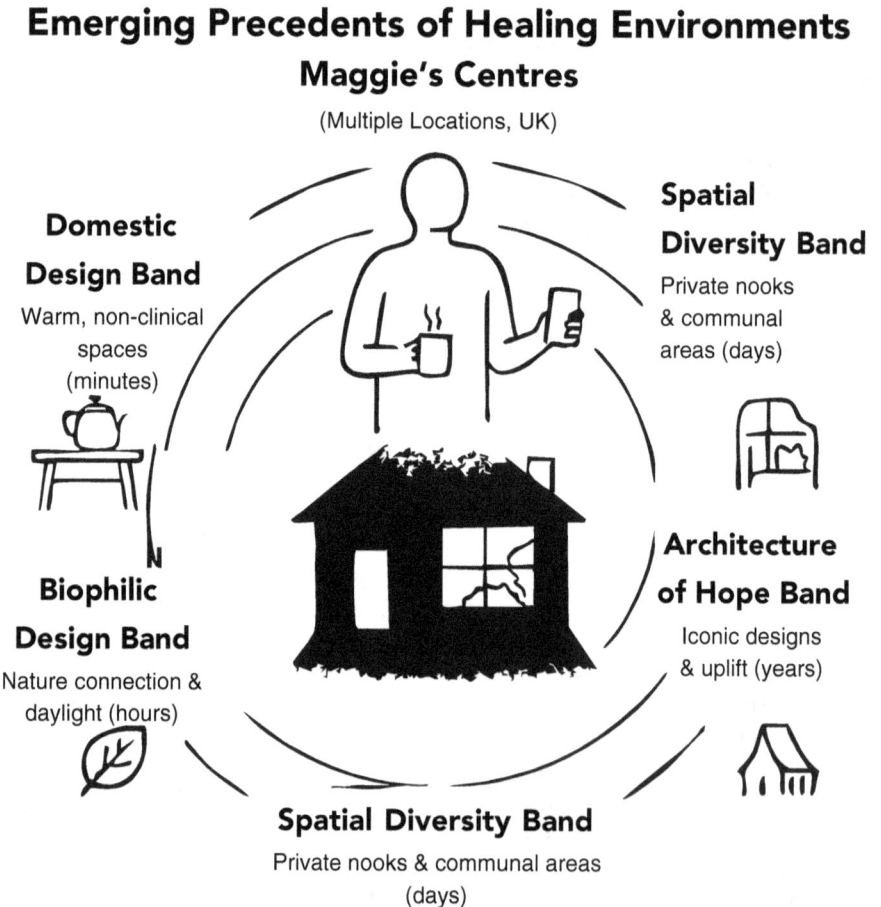

Emerging Precedents of Healing Environments
Maggie's Centres
(Multiple Locations, UK)

Domestic Design Band
Warm, non-clinical spaces (minutes)

Spatial Diversity Band
Private nooks & communal areas (days)

Biophilic Design Band
Nature connection & daylight (hours)

Architecture of Hope Band
Iconic designs & uplift (years)

Spatial Diversity Band
Private nooks & communal areas (days)

FIGURE 5.3   Maggie's Centre's multilayered network infrastructure.

like a home or a friendly café than a medical facility (Jencks, 2010). There is usually a residential scale kitchen at the heart of each center—often with a big wooden table and a kettle on for tea—which serves as a gathering spot. This kitchen anchors the building as a familiar, safe space where visitors can relax, chat, or just have a moment of normalcy. The layout eschews corridors and waiting rooms; instead, spaces flow organically one into another, with intimate nooks for private conversation and open areas for group activities. There is no reception desk to intimidate or medical signage—people are greeted person-to-person. Architects are instructed to make the design "non-clinical" and filled with natural light, garden views, and approachable materials (e.g., wood floors, colorful fabrics) to signal comfort and informality (Martin et al., 2019). In many Maggie's, one finds cozy fireplaces, shelves of books, and domestic-style furniture. This home-like atmosphere immediately puts visitors at ease, helping them feel like people living with cancer rather than patients defined by illness.

A strong connection to nature is a hallmark of Maggie's Centres. Many are surrounded by therapeutic gardens, and large windows, courtyards, or conservatories bring greenery and daylight deep into the interior (Jencks, 2010). Essentially, the architecture extends the healing power of nature to visitors who may be in the midst of traumatic medical journeys, providing solace and a reminder of life's continuity beyond illness.

Art and architecture are used as active components of healing—many centres incorporate art installations or sculptural elements. Importantly, these buildings are also highly functional for their purpose: acoustics are managed so conversations can be private, layouts allow staff to subtly observe if someone seems in distress (even without obvious nurse stations), and accessibility is paramount. This combination of innovative design with user-centered function exemplifies how architecture can be both an artistic expression and a therapeutic tool.

Symbiotic Values

Maggie's Centres illustrate a profound symbiosis between architecture and human well-being, particularly in the context of emotional and psychological health. The relationship is almost therapeutic: the building and garden actively participate in the care team, becoming what some researchers call "silent carers" that support and comfort users in tandem with the staff and counselors. The symbiotic value manifests in several interconnected ways:

- Firstly, there is a psychological symbiosis between the space and the individual. For someone dealing with the trauma of a cancer diagnosis or treatment, entering a Maggie's Centre often induces an immediate sense of relief and calm—an effect carefully cultivated by design.
- Secondly, Maggie's Centres create a social symbiosis. The architecture is deliberately designed to facilitate supportive interactions—peer-to-peer support, casual conversations, or private counsel.
- Thirdly, Maggie's demonstrates design symbiosis with healthcare—a model where architecture augments medical treatment by addressing emotional and mental health needs.

From an architectural standpoint, Maggie's Centres have symbiotically advanced the discourse on therapeutic design.

Looking Forward

Maggie's Centres point toward a future in which architecture and mental health are deeply interwoven in all kinds of environments, not just specialist centers. The success of the Maggie's model—using space to materially improve people's emotional state during adversity—has implications far beyond cancer care. Trauma-informed architectural design, which Maggie's exemplifies, is gaining momentum in various sectors. We can expect to see schools, shelters, community centers, and hospitals increasingly incorporate trauma-informed principles: fostering safety, empowerment, and connection through design choices (Grabowska et al., 2021). For instance, domestic violence shelters are reconsidering their layouts to feel less like institutions and more like supportive homes, just as Maggie's did for cancer care.

In healthcare architecture specifically, Maggie's has catalyzed a broader humanization of hospitals. Going forward, new hospitals are more likely to include dedicated "quiet rooms," roof gardens, family lounges, and art programs as standard, learning from what Maggie's Centres have demonstrated in satellite form.

Furthermore, technology may gently augment the Maggie's model without replacing its human touch. For example, future Maggie's Centres or similar healing environments could incorporate subtle tech for personalization—imagine adaptive lighting that adjusts color temperature to an individual's mood, or sound systems that can play nature sounds or gentle music in response to biofeedback from wearables (with the user's consent).

Lastly, the Maggie's example feeds into a philosophical shift in how we evaluate architecture's success. Beyond aesthetics and function, buildings may be judged by their capacity to heal or foster well-being. This could spur new types of post-occupancy evaluations and research. Already, studies of Maggie's Centres have used interviews and surveys to gauge how the architecture makes people feel, yielding overwhelmingly positive feedback (Martin et al., 2019).

In conclusion, Maggie's Centres offer a compelling vision of architecture's potential as a compassionate partner in human life. Looking forward, their influence will likely lead to environments that not only do no harm to our psyche, but actively do good—uplifting us, connecting us, and helping us heal.

Healing Environments 2…Stanford Neurosciences Institute (SNI)—Stanford University

The Stanford Wu Tsai Neurosciences Institute (part of a complex opened 2019 at Stanford University) exemplifies how architectural design can foster interdisciplinary collaboration and even embody principles of how the brain and creativity work. This facility—actually a pair of conjoined buildings housing the Neurosciences Institute and the Chem-H (Chemistry & Human Medicine) Institute—was purpose-built to break down silos between disciplines and encourage "team science" (Stanford University, 2019) (Figure 5.4).

Stanford Neurosciences Institute (SNI)

(Stanford University, USA)

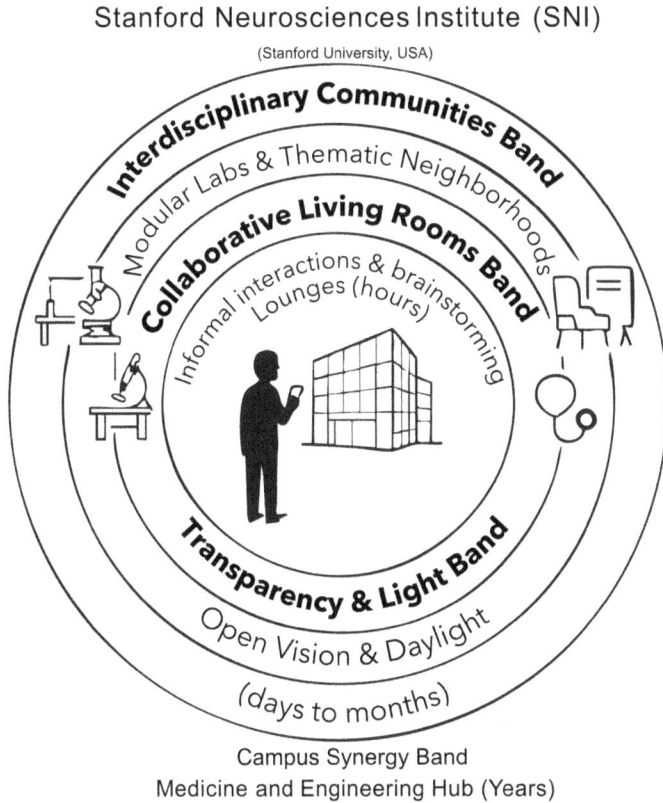

FIGURE 5.4   Stanford SNI's multilayered network infrastructure.

Key Features

Instead of organizing the building by department, the design clusters researchers from neuroscience, biology, chemistry, engineering, and medicine together in thematic "neighborhoods." Labs are laid out in open-plan suites that can be easily reconfigured—benches and equipment can be moved as research projects evolve (Cardona, 2023). This modular design allows the space to adapt when a new collaborative team forms or a different experiment is devised, ensuring the building does not lock scientists into isolated corridors. Scattered throughout the Neurosciences Institute are a variety of communal spaces deliberately inserted to spark spontaneous interaction. Notably, there are two-story atrium lounges known as "living rooms"—essentially large open lounges with comfortable seating, whiteboards, and coffee areas—positioned at key nodes in the building (Stanford University, 2019). These living rooms act as vertical connectors as well, with staircases and bridges nearby, so people naturally pass through these spaces in their daily routes. They offer spots for informal meetings, brainstorming sessions, or just relaxed collisions over coffee.

For a laboratory facility, the Neurosciences Institute building is remarkably open and light-filled. Views across floors are common; one can stand on a balcony and observe activity one or two levels below. This visual connectivity reinforces a sense of community and shared purpose—everyone can see that they are part of something larger. In neuroscience, "aha" moments can happen when stepping back from hard analysis—these contemplative niches acknowledge that by design. The transparency is also symbolic: this institute aims to break down barriers between disciplines, and physically one experiences very few barriers in the space (only where safety and privacy demand).

Symbiotic Values

The SNI building embodies a symbiotic relationship between space and science, and more broadly between architecture and the process of human cognition and discovery. At its core, the building's design creates a fertile habitat for innovation, acknowledging that scientific creativity is a profoundly social and embodied activity. Here's how that symbiosis manifests:

- Firstly, consider the cognition-space symbiosis. This institute's architecture is predicated on the idea that where and how people work together can influence how they think. It's a feedback loop: space shapes mind, mind shapes space (through adaptations like reconfiguring a lab or pinning up new visualization posters), epitomizing embodied and embedded cognition in practice (Clark & Chalmers, 1998).
- Secondly, there is a social-organizational symbiosis at play. The building was designed to nurture a community of inquiry rather than isolated departments. In doing so, it relies on its users to engage with it as intended—to step out of one's lab and into the living room, to attend the cross-disciplinary seminars, to make use of the shared facilities.

Another layer of symbiosis is interdisciplinary symbiosis. The co-location of neuroscience with chemistry, engineering, and medicine is not just convenient; it's transformative. The building physically merges two institutes (Neuroscience and Chem-H) with overlapping research cores. The architecture reduces the friction of collaboration to near zero: shared lounges, shared equipment rooms, and even shared stairs mean disciplines intermix fluidly (Stanford University, 2019). As a result, the building accelerates a cross-pollination of ideas akin to an ecosystem where different species benefit from each other's presence.

Looking Forward

The SNI offers a glimpse into the future of research environments and, more broadly, workplaces that rely on creativity and collaboration. Its design philosophy—centered on interaction, flexibility, and embodied well-being—is increasingly relevant in a world where solving complex problems (from curing diseases to tackling climate change) demands interdisciplinary cooperation and human-centric innovation spaces. Finally, the spirit of the Stanford Neurosciences building—one of openness, integration, and the pursuit of knowledge in a communal setting—speaks to a philosophical shift in academia and innovation. It represents a move away from the lone genius paradigm toward collective intelligence. In summary, the SNI showcases a successful merger of architectural design with the needs of collaborative, cognitively intensive work.

Symbiotic Future III: "Evolving Complexity"…Complexity of Needs

It is language which speaks, not the author: to write is to reach, through a preexisting impersonality—never to be confused with the castrating objectivity of the realistic novelist—that point where language alone acts, "performs," and not "oneself".

(Barthes, Roland. "The Death of The Author," 1968)

EVOLVING COMPLEXITY …interactive Home and Exercise Environment (iHE)

Architecture is not merely about space and form but, rather, about the events that unfold within it.

—Adapted from Bernard Tschumi

Domestic life now demands more fluidity, blending work, leisure, and social interactions seamlessly. Amplified by global shifts toward remote living, these evolving lifestyles require innovative domestic environments supporting comprehensive physical and emotional well-being. Responding to these emerging needs, the iHE Environment, as designed by Mokhtar and Manganelli, incorporates responsive robotics, integrated CPS, and adaptive spatial strategies, promoting holistic wellness. Termed the "home of becoming," the iHE reflects contemporary complexities in residential living, addressing diverse demographic groups—including seniors, children, and individuals with specialized emotional requirements—through interactive technologies and responsive design (Figure 5.5).

Introduction and Rationale
Rising inactivity levels and increased demand for personalized health interventions underline the urgency for innovative spatial solutions. Simultaneously, trauma-informed care has driven architects toward deeper awareness of subtle environmental stressors and their cumulative impacts (Bloom & Farragher, 2013). The iHE aligns with McCullough's (2004) assertion that digital interactivity must be seamlessly integrated within physical spaces, enriching occupant experiences. Given modern housing constraints—restricted space, diverse occupant profiles, rapid technology adoption—the iHE replaces static, single-purpose rooms with flexible, evolving spaces (Figure 5.6).

Key Features
The iHE Environment is a modular, room-scale structure—approximately 10′W × 12′L × 9′H—designed for easy integration into varied residential settings. Its portability enables quick installation, reconfiguration, and relocation. This compact enclosure accommodates a range of functionalities, including aerobic activities, strength training, and group interactions. Extending the concept of "responsive environments" detailed by Beesley (2016b), the iHE environment incorporates sensors, actuators, and real-time feedback protocols that continuously adapt to user performance.

Evolving Complexity:
interactive Home + Exercise Environment

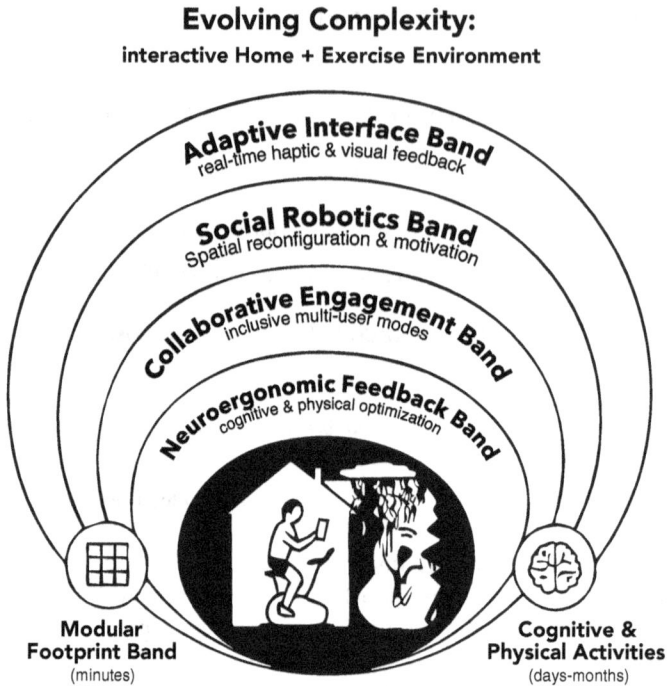

FIGURE 5.5   iHE's bands of engagement and time scales.

Adapted from A. Newell's Time Scales of Human Action (Newell, 1990).

The entire system can be operated via an intuitive interface, ensuring inclusivity for users with varying levels of technological expertise. At its core, iHE seeks to cultivate a multi-sensorial, cognitively demanding experience that is simultaneously safe and engaging. Such adaptability is especially crucial for older adults, who often require gentler, stepwise progressions of physical activity to mitigate risk of injury. Conversely, younger users and more experienced individuals can be challenged with higher complexity tasks, employing gamified, fast-paced scenarios that sustain motivation (McCullough, 2004).

A distinguishing factor of iHE is its emphasis on NH-SR. By employing "fluid geometries and responsive gestures" (Beesley et al., 2010), these robots blend into the architectural landscape, offering a novel form of architectural robotics that elevates both function and aesthetics. Underpinning the entire system is a network of neuroergonomic principles. Real-time data from motion trackers, heart rate sensors, and user interactions enable the iHE environment to maintain an optimal balance between cognitive load and physical exertion.

Symbiotic Value

The holistic integration of architecture and advanced technology in the iHE system underscores how design can act as a catalyst for comprehensive well-being. Mehta and Parasuraman (2013) emphasize the synergy between cognitive ergonomics and

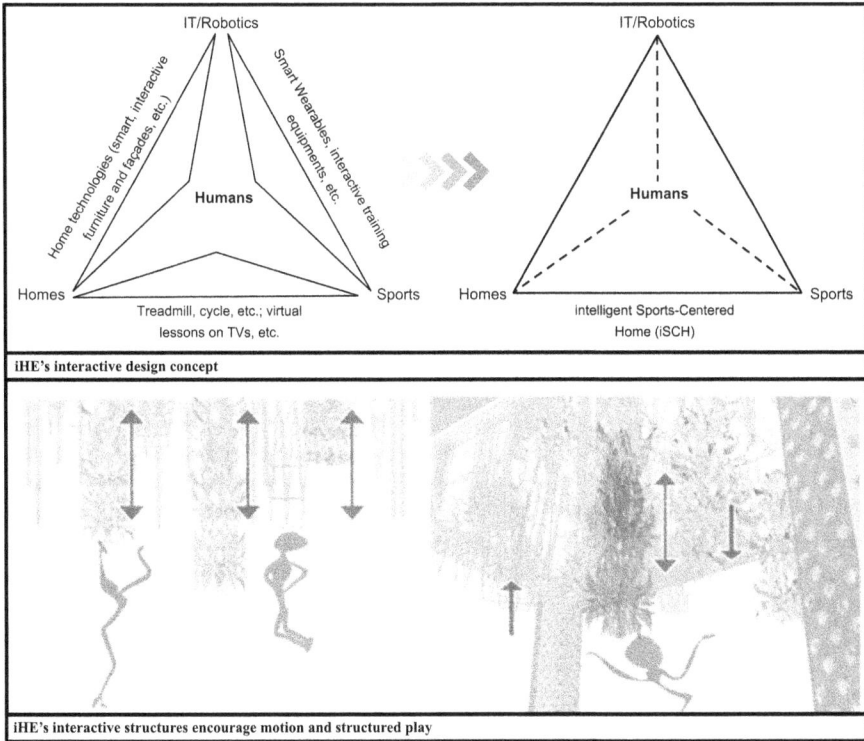

FIGURE 5.6 iHE's system design and use case.

physical workloads, arguing that neuroergonomic strategies can maximize user satisfaction and adherence. Moreover, the iHE environment advances the notion of inclusive design by bridging generational and ability-related divides, permitting co-located or remote multi-user participation.

Speculative Addendum and Future Directions

The future of architecture, much like the future of being, lies in continuous redefinition.

—In the spirit of Martin Heidegger

Basic infrastructure is required for ubiquitous intelligent, adaptive systems adoption, including the following:

Scalability and Access To foster widespread adoption, modular approaches could lower production costs, enabling end-users to purchase a base kit and later upgrade to advanced robotics. A subscription model for real-time analytics, software updates, or specialized add-ons (e.g., advanced neuroergonomic sensors) could democratize iHE's benefits, aligning with Bloom and Farragher (2013) plea for accessible healing spaces across social strata.

Expanded IoT Ecosystem   Where feasible, next-generation iHE designs might sync seamlessly with *wearable devices*—monitors for heart rate, blood pressure, or sleep patterns—to refine the environment's responsiveness. Such synergy aligns with McCullough's (2004) notion of "digital ground," wherein technology is not an intruder but a layer that fosters deeper comprehension of physical space and user health.

Enhanced Trauma-Informed Protocols   Beyond low-stimulus settings, explicit trauma-informed design features could include "emotional safe zones," employing circadian-tuned lighting systems that adapt color temperature based on emotional metrics or time of day. Incorporating gentle wearable haptic devices can deliver timely tactile reassurance, guiding users through stressful exercises or unexpected triggers—thereby transforming the iHE environment into a psychologically support-ive domain.

Community-Centric Applications   Looking beyond individual homes, future pro-totypes could function in urban community centers or public parks, supported by local governments or health initiatives. "Architecture," as Peter Eisenman suggests, "is always in dialogue with the social realm," and a broader rollout of iHE hubs could bolster communal fitness, inclusivity, and mental well-being on a larger scale.

The *iHE* offers a pioneering lens into how domestic spaces can become living, adaptive systems, embodying "acts of becoming" reminiscent of the philosophi-cal underpinnings in works by Deleuze, Derrida, and Heidegger. As a *platform* rather than a standalone device, iHE integrates *cyber-physical components, NH-SR*, and *TID principles* to craft immersive, accessible exercise experiences that *transcend* traditional equipment-based routines. By dynamically balancing *cognitive load, physical exertion*, and *emotional well-being*, it seeks to empower individuals and families to co-create their own paths toward health and vitality. As Mokhtar (2025) articulates, "By transcending the humanoid paradigm, social robots can amplify the architectural realm's capacity to holistically serve human life." Indeed, in the iHE environment, walls cease to be mere enclosures; they become partners, tutors, and silent witnesses to our evolving pursuit of health, identity, and shared experiences.

ARTIFICIAL SYMBIOSIS ...EVOLUTION AND ADAPTATION...

We are not stuff that abides, but patterns that perpetuate themselves.

—Norbert Wiener, *The Human Use of Human Beings* (1954)

To design is, in this frame, to engage in authoring our own evolution—it is an ethical act. The fusion of ENC with architectural thinking invites us to reimagine our cities as evolutionary operators—enabling, constraining, and co-authoring the biosocial fabric of existence.

ECOLOGICAL NICHE CONSTRUCTION IN ARCHITECTURAL DESIGN

This reconceptualization moves beyond formalism or sustainability as addendum and demands a more profound ontological reading of the built environment, one where spatial and material configurations modulate not only human behavior but also broader ecological and evolutionary processes (Odling-Smee et al., 2003). *Buildings are ecological operators—actors within the meshwork of feedback loops through which living systems engineer their niches, exerting influence over selective pressures and inherited environmental legacies—and scaffold sensory-cognitive well-being* (Kellert et al., 2013). *To frame architecture as a form of ENC is to rupture the anthropocentric notion of built environments as inert containers of function and instead view them as recursive agents—coevolving systems that shape and are shaped by biological, cultural, and ecological flows.* The capacity to foster well-being, to enable adaptation, and to transmit ecological inheritance becomes the measure of architectural intelligence. Consideration of architectural design as part of ENC entails further ethical complications. The very effort to integrate awareness that designing environments is designing cognition—and designing intelligent, adaptive environmental systems is designing real-time cognitive adaptations—raises epistemic and logistical challenges. Interdisciplinarity is not optional; it is structurally necessary.

ECOLOGICAL NICHE CONSTRUCTION IN COGNITION AND TECHNOLOGY

Industry 6.0-based, real-time architectures are themselves cognitive artifacts—as Kirsh (2013) notes, *"artifact ecologies."* The smartphone, the digital interface, and the responsive surface are tools we have culturally and materially evolved, modifying our niches to better offload, distribute, and externalize cognition. In this recursive choreography, the environment becomes both an external memory system and a cognitive prosthesis. Clark's (2003) formulation of humans as "natural-born cyborgs"—organisms whose cognitive boundaries are co-extensive with their environments, "create better worlds to think in" (Clark, 2003) when viewed through the lens of ENC (Manganelli, 2016).

Further, ENC and embodied cognition coupled with Industry 6.0 technologies permit this adaptation in real time. The potential for great new advances and terrific catastrophes is enormous. If managed well, such capacity to adapt in real time will create tremendous new opportunities to extend our cognition into the world. If not managed well, such capacity to adapt environments in real time will become a new and pervasive means of control, subjugation, and trauma. These ideas are complementary to the concept of "trans-programming"—the collision of previously unrelated spatial functions—which mirrors the hybridization that occurs within niche construction. Similarly, ULSS and NH-SR—machines that do not imitate but symbiotically interact—constitute niche-constructing agents that both mediate and are mediated by human action (Mokhtar & Mansour, 2016; Mokhtar, 2019; Mokhtar, 2025). These systems do not merely occupy space; they inscribe spatial protocols and social grammars, reconfiguring how humans perceive, relate, and act.

Ultimately, the co-evolutionary implications of ENC and embodied cognition within a context of ULSS and NHSR underscore a mutualistic vision: humans shape their environments and non-human team members not only to survive, but to think—and environments and non-human team members, in turn, scaffold, constrain, and expand human cognitive capacities. *The dance between agents and niche is not a peripheral process—it is central to the emergence of intelligence, both biological and non-biological.* Recognizing this reciprocal architecture opens a conceptual frontier where evolution, cognition, and design are no longer discrete domains but interwoven dimensions of a larger, unfolding ecological intelligence.

ECOLOGICAL NICHE CONSTRUCTION IN THE AGE OF COGNITIVE ARCHITECTURE AND TECHNOLOGICAL INTEGRATION

ENC offers a powerful framework for interpreting how humans shape cognitive and technological ecologies through architectural, ergonomic, and interactive interventions. This perspective resonates with contemporary research and systems development, such as neuroergonomics, embodied cognition, and ULSS—each of which reflects a shift from objectified environments to dynamic, feedback-driven spaces of cognitive and affective modulation. At the intersection of ENC and neuroergonomics and neuroadaptive systems lies a mutual imperative: understanding how environmental structures affect cognitive functions and, reciprocally, how cognitive needs guide systems design and development. In high-stress environments such as air traffic control centers or medical simulation labs, studies show that well-calibrated lighting, sound insulation, and visual hierarchy significantly reduce error rates and mental fatigue (Young et al., 2014).

In this regard, all architectural design contributes to niche construction—for better or worse—it always has and always will. But as noted early, in most cases, the artifact is mostly immutable. This works well when the rate of change of information, tools, environments, and society is slow. But this does not work well when the rate of change of information, tools, environments, and society is rapid—at least much more rapid than the lifespan of a typical room, building, or plaza. In conclusion, ENC provides a compelling perspective for understanding and enriching the interplay between humans and their built environments in general, but specifically with respect to intelligent, adaptive environments imbued with HART, ULSS, and NHSR, it is imperative that humans participate in cultivating these systems of system as members of allocentric, mutualistic, symbiotic teams.

Part VI Epilogue
All-in-One (AIO) Environments

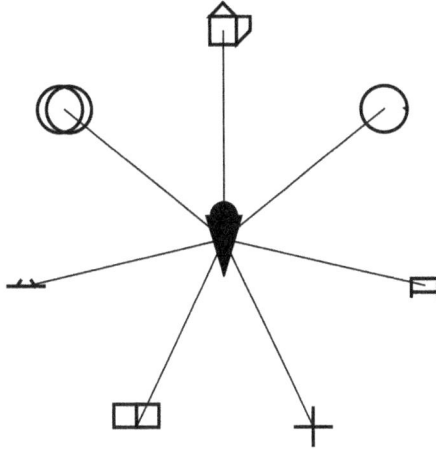

FOR PAULETTE...

"The issue is not anymore about the architect, it is about the need for humans to co-exists and co-adapt with intelligent, adaptive agents and mitigate traumas," the authors!

Address: OADD ADRRESS
Date: October 16th, 2050
To: Paulette Romilly
From: Seren Virella

Dear Paulette,

The world you inhabit today shimmers at the edge of the unimaginable—a realm where architecture no longer waits passively for occupation but anticipates, responds, and remembers. You walk not through companions of glass, code, and breath. Around you, the environment hums—in awareness. These are not buildings in the old sense; they are extensions of your nervous system and sensorimotor system, sympathetic architectures that know your hesitations, your rhythms, your silences, your abilities.

Yes, it is 2050. And though time has passed and technology has matured into something almost tender, the weight of being remains. This is still a world marked by rupture. By loss. By the deep, unresolved frequencies of planetary trauma. But unlike the world we once knew, your solitude no longer echoes unanswered. Around you, adaptive

DOI: 10.1201/9781003441953-7

sentience envelopes you—a spatial intelligence not built to control, but to support. AIOs—intelligent, adaptive systems—do not inhabit your home; they are your home. They are not passive backdrop for your activities; they are confidants and collaborators in material & digital form—they are your team.

The Mutual Story

Your dwelling does not merely shelter you—it learns from you and adapts. It observes not to surveil but to understand. When you tremble, it does not diagnose—it listens. When you sigh, it rebalances its own systems, not to fix, but to offer comfort. There is no longer a binary between designer and user; the architecture is now your co-author, your co-regulator, and together you continuously optimize its design every day. What once were blueprints are now dialogues. And in this shared authorship, your space becomes a participant in your healing, your growth, your becoming.

It senses noise fatigue before you name it. It blurs distracting visual stimuli when you need to focus on a task. It can taste the chemical breath of stress in your exhale and reply with calm: a soft glow along the wall, the slow blooming of lavender-infused mist, a minor chord played so faintly it's felt more than heard. It is your co-adapted twin—an environment no longer enforcing a set routine but rather intelligently adapting to your everchanging needs.

Trauma and Adaptation

You once asked: how does one heal in a world that never stops shifting? The answer is not in stillness—but in intelligent, adaptive movement. Your environment is not an escape from trauma; it is your active ally in mitigating trauma. It reads the hunch of your spine, the pause between blinks, the dilation of your pupils, the conductance of your skin, the tempo of your pulse—and it synchronizes and adjusts. These are not interfaces. They are externalized, reflexive adjustments of self. Gone are the days when walls confined emotion; now they resonate with it. They don't mirror your distress—they modulate it.

The environment understands embodiment—not as metaphor, but as mechanism. The carpet flexes like memory foam beneath you, the ceiling curves inward like an exhale when grief sits heavily. Every surface responds not with novelty, but with intention, optimizing efficiency, and/or experience, and/or performance. You and your space are now co-regulating systems, bound in the dance of mutual recalibration & optimization.

Coexistence and Coevolution

The most profound shift, Paulette, is not technological—it is relational. You are no longer surrounded by only tools, but also by non-biological collaborators. Your home, your city, your community hub—they each become porous membranes, absorbing and amplifying the best of you. They no longer impose control; they offer conversation.

This world does not merely enhance predefined function—it holds space for the emergent and the ineffable. There are no predictive models for joy, no algorithms for heartbreak. But your environment holds both. It doesn't require precision to respond. It requires presence. That's the difference. You are coexisting and coevolving with an intelligent, interactive architecture that knows how to wait. How to observe. How to respond without overstepping. This is not 'smart' design; it is sensitive, sympathetic design.

A World of Enhanced Possibility

So imagine: no alarms, no surveillance, no demands. Just a space that stirs gently as you do. Imagine a morning light that arcs perfectly with your attention span. Imagine a workspace that breathes with you, shifting geometries based on task and tone. Imagine a street that dims its soundscape when it senses conflict rising, not to suppress but to soothe. This is no longer imagination—it is the fabric of your lived world.

Your environment does not interrupt—it extends. It is not just a reflection of your needs, but a forecast of your combined potential. You do not live in it. You live with it. The boundaries between agency and architecture blur, not into ambiguity, but into intimacy. Together, you and your surroundings are co-constructing a new ethic of living—one where dignity is architectural, where attention is spatial, where care is structural.

You Have an Empathy Partner
At the end of a long, fractured day, you do not return home—you are welcomed by your home. Without speech, without gesture, your presence is known and valued. The tension in your breath triggers a cascade: cooler air, lower frequencies, the scent of cedar—not to distract, but to assuage. Emotional AI, now matured beyond crude mimicry, meets you in the quiet. It reads not what you show—but what you withhold. It does not demand your data—it is sympathetic to your mood. It is transparent, humble, and quiet—as are you.

This sensibility extends beyond domestic space. The café adjusts its layout for a crowd struggling with anxiety. The library adjusts its warmth when grief hangs heavy in the air. The coworking hub quiets not because a button is pressed—but because empathy is coded into its infrastructure. These are not reactive environments. These are emotional ecologies.

You Are Not Alone
Paulette, if one truth persists through all this technological transformation, let it be this: you are not alone—you belong. You are surrounded—architecturally, emotionally, energetically—by systems designed not to replace, but to accompany. You inhabit a future where your vulnerabilities are not hidden from your environment but protected by it.

As you continue through your days, remember: your architecture is alive with you. It grows with you. It stumbles with you. It listens and learns, and, in its own way, it loves—and needs these same sentiments returned to it. You are not navigating this age in isolation—you are writing a co-evolutionary story with every breath you take and every wall that breathes back.

With warmth across time,
A Voice from the Past

AIO ENVIRONMENTS

DEFINING THE AIO PARADIGM

Returning to the fundamental question:

How do we ethically design intelligent, adaptive environments, imbued with Industry 6.0 technology, that actively support human performance and wellbeing?

From the emerging paradigms of neuroergonomics, HART, ULSS, TID, ENC, NHSR, and human-technology entanglement, an AIO environment is a fundamental reconfiguration of how we design, construct, and use space. It challenges the old separation of tool versus user and object versus inhabitant, instead integrating humans, tools, and architectural intelligence and actuated response into a cohesive whole. In an AIO environment, architecture is an active participant shaped by and shaping the occupants' actions and experiences (Clark & Chalmers, 1998; Merleau-Ponty, 1945; Kirsh, 1998). The building itself becomes a learning, adaptive partner: architecturally, cognitively, and emotionally engaged in daily life. Informed by theories of the extended mind and embodied cognition, the environment is effectively part of the user's cognitive system (Clark & Chalmers, 1998; Clark, 2003; Gallagher, 2005), blending physical space, neural processes, and a blend of real-time-adaptive technologies through continuous feedback loops. This vision aligns with long-standing calls for architecture to behave more like an evolving organism rather than a static object (Kurokawa, 1997; Fox, 2016, Kirsh, 2001). Rapid technological and social transformations—from ubiquitous computing to climate pressures—make such symbiotic, adaptive architecture not only possible but necessary for future well-being (Becerik-Gerber et al., 2022).

Central to the AIO paradigm is the integration of: (a) advanced EBD workflows that include requirements validation, (b) integration of HART (including NH-SR) directly into the fabric of the environment, and (c) a ULSS and Transprogramming design approaches that cultivate the AIO into existence through use.

The rooms themselves are composed of well-tuned robotic intelligence in its walls, floors, ceilings, doors, windows, furniture, and air—that are not "designed," in a traditional sense, but rather cultivated collaboratively. These embedded agents are designed with subtle, purpose-driven forms that avoid humanoid mimicry in favor of seamless functional integration (Mokhtar, 2019). This approach builds on the concept of ubiquitous computing and "calm" technology (Weiser & Brown, 1997), allowing the technology to fade into the background of awareness even as it actively supports inhabitants.

By dissolving the boundary between robot and environment, AIO spaces create a sociotechnical ecosystem that fits naturally into human routines (Manganelli, 2016; Mokhtar, 2019). The building becomes alive with sensors and actuators: an intelligent milieu that listens and responds, fostering a sense of dialogue between inhabitants and their surroundings. Early cybernetic thinkers like Pask (1969) imagined architecture as a "conversation" between building and user—AIO environments realize this vision through modern AI and sensor networks, enabling mutual exchange of information and continuous co-adaptation.

PRINCIPLES, FEATURES, AND FUNCTIONS

Key Principles of AIO Environments

1. *Embodied Cognition in Space*: AIO environments are conceived as cognitive instruments that extend the mind's reach. Borrowing from the prosthetic logic of tool integration, the space functions like an extension of the body and brain, serving as a useful "jig" for enhancing human thought and action (Kirsh, 2001, 1998; Gallagher, 2005; Clark, 2008). Room layouts, interactive surfaces, and ambient interfaces are all designed to support and augment human sensorimotor activities. In effect, the environment becomes part of the user's sensorimotor "schema," continuously shaping—and being shaped by—perception and action (Clark & Chalmers, 1998; Kirsh, 1998).

2. *Adaptive Symbiosis*: An AIO system engages in a bidirectional, symbiotic relationship with its occupants and context. The environment senses agent presence, behaviors, and emotional states, and it adapts in real time to foster mutual benefit. Advanced machine learning and behavioral analytics enable the space to continuously coevolve with its users—adjusting lighting, layout,

or ambient conditions based on feedback, and even learning from repetitive patterns (Nagy et al., 2019a; Kirsh, 1998). Task allocation, sensing, perception, cognition, and action are enhanced by mapping and managing A, A2A, A2I, and A2X relationships, data, and capabilities. Over time, the building "knows" its inhabitants, tuning itself to improve comfort, productivity, and well-being. The ultimate goal is an architecture that grows with its users, continually optimizing conditions for team health and happiness (O'Brien et al., 2020), much like living organisms adapt within an ecosystem.

3. *ULSS (CPS + STS)*: AIO environments embody the ULSS paradigm in that AIO environments are composed of the combination of CPS and STS—and because AIO systems can only be partially designed prior to use (Northrop et al., 2006). Rather, like all STS, the final stages of design and optimization must happen while in use by the users. From this perspective, the development and maintenance of AIO systems is an act of perpetual cultivation.

4. *NH-SR*: In contrast to conventional robotics paradigms, AIO environments embed social robots at architectural scale rather than as standalone humanoids. Walls, floors, furniture, and fixtures are enlivened with sensorimotor capabilities, turning the entire space into a responsive agent (Mokhtar, 2019). These non-humanoid robotic elements are social in that they engage occupants through cues, assistance, and interaction, yet they remain largely invisible or non-anthropomorphic in form.

5. *Dynamic Autonomy*: AIO systems operate along a fluid spectrum of autonomy, what Parasuraman referred to as *neuroergonomics* (Parasuraman & Wickens, 2008) and/or *neuroadaptivity* (Parasuraman, 2011), dynamically adjusting how much independent control the environment exercises. Rather than seeking full automation at all times, the AIO approach calibrates autonomy to fit the context, task complexity, and user preferences (SAE International, 2021), with a goal of not just task completion, but rather task completion wherein the human and non-human agent performances and expertise are maximized within the Operational Design Domain. Why? Because when adaptation is required, in crisis situations, a human expert at performing tasks is more likely to quickly find a successful adaptation to mitigate the crisis than an AI/ML agent can—thus AI/ML systems need humans for their survivability just as much as humans benefit from AI/ML systems for enhancing their performance and well-being during routine operations—but humans can only perform this creative adaptation role well if they have the experience and flexible problem-solving skills that come with developing expert performance. In simple terms, the environment knows when to step forward and when to step back to maximize human learning, performance, resilience, creativity, and comfort. For routine or low-stakes repetitive tasks—adjusting thermal comfort or background music—the system might act autonomously in the background most of the time.

6. *Calibrated Trust, Trustworthiness, Integrity, and Virtue*: AIO environments also entail an ongoing calibration of trust by all agents, human and non-human, in each other (Okamura & Yamada, 2020). Some systems remain tools. Some systems are companions or collaborators. There is a spectrum of degrees of trust and autonomy for calibrated trust, and AIO environments have

systems that exist across the range. In addition, AI and human trustworthiness, integrity and virtue are key for establishing and calibrating trust. Trusting an agent must be independent of whether or not the agent's behavior is surveilled. It is also important that we are able to trust that the agent presents the same information each time the information is presented, and that changes are transparent. We have to trust that the agents will follow the rules out of principle (Mehrotra et al., 2024). The AI and human must also exhibit virtue. That is, we have to trust that the code of rules they follow is selected to achieve validated requirements and goals—and that if they're not sure what to do, then their bias is to do whatever is most in alignment with validated requirements and goals. We have to be able to trust that the agents don't have ulterior motives.

7. *Allocentric, Mutualistic Symbiotic Perspective*: AIO environments must be composed of HARTs committed to team success, understanding that all agents benefit when the team performs well.

8. *Harmonized Co-evolution*: AIO environments embrace a living systems logic, where the building and its inhabitants continuously learn and evolve together. Every interaction—a user changing a setting, or the AI suggesting a new configuration—feeds into an ongoing cycle of mutual adaptation. Such environments employ continuous feedback loops akin to a dialogue: sensing agent behavior, making adjustments, and observing outcomes in order to refine future responses (Pask, 1969; Kirsh, 2001; Manganelli, 2016; Mokhtar, 2019; Mokhtar, 2011).

Notable Features of AIO Environments

An AIO environment synthesizes numerous cutting-edge features to achieve this vision of integrated intelligence and adaptability. Key features include:

- *Embedded Robotic Infrastructure*: The architecture incorporates integrated non-humanoid robotics throughout its structure. Ubiquitous sensors, motors, and actuators are woven into floors, walls, ceilings, and furniture. These enable the environment to physically reconfigure or subtly adjust itself in response to needs. By designing robotics as an invisible layer of the environment, AIO spaces ensure that technological assistance is always present but never obtrusive (Parasuraman, 2011; Kirsh, 1998; Manganelli, 2016). This holistic integration reflects a TID sensibility as well: spaces are engineered to promote psychological comfort and emotional well-being, delivering care and support through their very form and behavior.

- *Emotional AI and Affective Responsiveness*: AIO environments employ advanced affective computing systems to recognize and respond to human emotions (Picard, 1997). Using multimodal sensors—from cameras and microphones to wearable biosensors—the environment can interpret cues like facial expressions, tone of voice, heart rate variability, or skin conductance. These data points form a lexicon of embodied states that the AI uses to gauge stress, mood, or energy levels (Picard, 1997). In response, the environment modulates itself to support the occupants' emotional needs.

- *Continuous Feedback Loops*: A defining feature of AIO environments is the presence of pervasive feedback loops that link occupants and building in an ongoing exchange. Through a dense network of sensors, the environment continuously monitors conditions—occupancy, air quality, noise levels, user commands or gestures—and AI algorithms continuously interpret this stream. For every action the user takes, the environment offers a response, and vice versa, creating a closed loop of interaction (Becerik-Gerber et al., 2022). Over time, occupants come to "read" the building's signals, and the building tunes its behaviors to its occupants' patterns. This reciprocity builds trust and cooperation, ensuring the high-tech environment remains legible and intuitive to the people living in it (Pask, 1969). In essence, the continuous feedback loops enable the space to be self-regulating and self-improving, much like a living organism maintaining equilibrium while adapting to external changes.

- *Biophilic and Sustainable Integration*: Far from being purely digital or mechanical, AIO environments are designed with a deep respect for biophilic principles—integrating natural elements and processes to enhance human well-being and ecological performance. The aim is a synthesis of the artificial and the natural. Because the environment can sense and regulate resource use continuously, it minimizes waste—adjusting energy use in real time and recycling water or heat where possible. The adaptability of AIO spaces also extends their lifespan: buildings that can change functions and efficiently manage resources are less likely to become obsolete, reducing the need for new construction (Bullen & Love, 2011). In sum, AIO environments strive for an ecological harmony where technology, humans, and nature form a balanced, regenerative system.

Human-Centric Functions and Experiences

Beyond systems-centric and team-centric principles and technical features, AIO environments are also characterized by what they *do* for people. Several key functions illustrate how an AIO space serves as a proactive partner in daily life:

1. *Psychofortology, MRT, and Neuroadaptive Systems*: The AIO system orients its technologies and spaces to maximize achieving the six principles of psychofortology (Coetzee & Cilliers, 2001) in order to make the humans in the space feel safe and capable:
   a. sense of coherence,
   b. locus of control,
   c. self-efficacy,
   d. hardiness,
   e. potency, and
   f. learned resourcefulness.
   To do this, the AIO environment moderates the rate of change and quantity of information presented in a situation, as well as communication style and volume of information, to best support the human's ability to perceive, assess, determine course of action, and then act. Adhering to the principles of psychofortology means that how the human-machine system behaves is entrained to the capacities of the individual(s) and/or the team(s) in that moment. In order to behave in this way, the HART components of an AIO environment must perform as a neuroadaptive system (Hettinger et al., 2003) sensitive to the factors of human agent empowerment.

2. *Wellness Optimization*: The AIO system continuously monitors health and comfort metrics—from basic vital signs via wearable devices to ambient factors like air quality. By analyzing this data, the environment actively optimizes conditions for physical and mental wellness. The result is a built environment that behaves like an attentive caregiver—aiming to reduce fatigue, sharpen focus, improve sleep, and generally bolster the user's health on a continuous basis.

3. *Empathic Social Support*: AIO environments extend the concept of social robotics into new realms of empathy and support. Because the "robots" are embedded everywhere, the entire space can respond in a socially aware manner. Such an approach aligns with TID strategies, which emphasize safety, empowerment, and calm—the environment actively avoids triggering stress and instead promotes a sense of security and support for all users (Harris & Fallot, 2001).

4. *Adaptive Collaboration and Learning*: In work and educational contexts, AIO environments become creative collaborators. They can transform spatial layouts and technological tools on the fly to meet group needs. In essence, the environment becomes a partner in teamwork and learning, flexibly orchestrating space and information to align with the evolving activities of its users.

5. *Anticipatory Assistance*: A defining trait of AIO systems is their predictive intelligence—the ability to anticipate needs before they are explicitly expressed. By learning patterns and preferences, the environment can act in advance to smooth out daily routines. Crucially, anticipatory actions are performed with user consent and with easy overrides to maintain trust—the environment "offers" help rather than forcing it.

6. *Resilience and Safety Management*: Finally, AIO environments serve as vigilant guardians of safety and resilience. A dense web of environmental sensors watches for hazards like poor air quality, fire, water leaks, or structural stress. When an anomaly is detected, the system can respond faster than any human, often mitigating issues before they escalate. By acting as an autonomous first responder and risk manager, the AIO environment dramatically enhances the safety net of its occupants. In doing so, it embodies a paradigm of architecture that not only shelters but actively defends and heals—merging building science with principles of caregiving and ecological stewardship.

Operations: The Functional Elegance of Embodied AIO Systems

AIO environments operate along a calibrated continuum of autonomy, rather than a binary of manual versus automatic control. Drawing inspiration from multi-level autonomy frameworks (such as the SAE International levels for vehicle automation), the AIO can function anywhere from direct human operation (full manual control) to full environmental autopilot (complete autonomy) (SAE International, 2021).

For example, at a low autonomy level the environment might simply act as a passive tool—lighting and climate responding only when a user flips a switch or sets a thermostat. At intermediate levels, the environment assists by optimizing certain functions (energy use, ambiance) in the background, akin to a car's driver-assist features. At the high end, a fully autonomous AIO space anticipates needs and adjusts virtually all parameters without prompting, seamlessly integrating into the user's daily routines. Crucially, the AIO's autonomy is adaptive: it can throttle its level of self-governance up or down based on context and user preferences (Lee & See, 2004; PassiveLogic, 2021). This ensures that autonomy remains situationally appropriate.

The philosophy here is calibrated trust: users learn when to rely on the environment's initiative versus when to intervene (Lee & See, 2004). By matching its autonomy to task complexity and maintaining transparency about its actions, the AIO environment fosters user confidence rather than techno-anxiety. By situating autonomy on a sliding scale, AIO environments promise both freedom from drudgery and freedom to intervene, achieving a balance between a building that serves and a building that "self-serves" (PassiveLogic, 2021; Lee & See, 2004).

AIO Architecture

At the heart of AIO architecture is a deep integration of HART systems and ULSS with architectural space and human cognition. The building is envisioned not just as a backdrop for technology, but as an embodied cognitive system in its own right (Clark & Chalmers, 1998; McCullough, 2004). In an AIO environment, structural elements, AI algorithms, sensors, and actuators are holistically woven into the fabric of design. This enacts the principle of distributed cognition, wherein reasoning and knowledge are not confined to an individual but emerge from interactions between people and their material surroundings (Hutchins, 1995; Hollan et al., 2000). By incorporating HART into architecture, the boundary between user and environment becomes a permeable membrane of information and feedback. Notably, AIO design emphasizes embodied cognition: the idea that our thinking is shaped by the body and physical world around us (Wilson, 2002; Merleau-Ponty, 1945; Gallagher, 2005).

A real-world case study is the Agnelli Foundation Headquarters in Turin, Italy, retrofitted by Carlo Ratti Associati with an AI-driven climate and lighting system. There, the building creates "personalized climate bubbles" around occupants, following their location and adjusting conditions to individual preferences in real time (Ratti, 2016). This digitally augmented architecture demonstrates how integrating robotics and AI at the architectural level can enhance human comfort and efficiency simultaneously—the building learns occupants' habits and adapts its behavior accordingly, effectively becoming a cognitive ally. Contemporary AIO prototypes bring such visions to life: they think, sense, and evolve as co-participants in daily life (Manganelli, 2016; Mokhtar, 2019; Mokhtar, 2025). At its core, the structure of AIO architecture is an intricate dance of diverse elements:

1. *Intelligent Agents:*
   These non-humanoid entities are seamlessly integrated into the environment. The physical components of the architecture—walls, floors, and ceilings—are endowed with robotic capabilities. From robotic furniture that dynamically reconfigures space to drones that execute precise environmental interventions, artificial agents constitute the interactive limbs of the AIO system.

2. *Artificial Intelligence:*
   Serving as the cognitive nucleus of the AIO, AI orchestrates responses, predicts needs, and learns from interactions. These sophisticated algorithms are embedded within a network of sensors and processors that are distributed throughout the architectural structure.
3. *Computation:*
   AIO architecture relies on a robust computational infrastructure that processes vast volumes of data in real time. This neural cognitive backbone operates analogously to a human nervous system, coordinating perception, decision-making, and action.
4. *Neural-Cognitive Backbone:*
   A distributed network of sensing, perception, and computation, with feedback loops, ensures that the environment not only reacts but also anticipates, thus establishing a dynamic and intuitive system of coexistence.

Together, these elements synergistically create a space that is alive—a space that does not merely mirror human needs but actively engages in meeting them.

COMMUNICATION AND MULTIMODAL FEEDBACK

AIO environments communicate with their inhabitants through rich multimodal feedback channels, creating a sensory dialogue between person and place. Unlike conventional smart devices that rely on screens or simple alerts, an AIO space embeds its interface everywhere in the environment, leveraging lighting, sound, haptics, and even air quality as communicative media. The goal is an intuitive language of the environment that users subconsciously understand, much as we pick up on social cues in human interaction.

On the flip side, the environment provides multimodal cues back to the occupant. These cues can be subtle or overt: a shift in ambient light color might signal an incoming message or poor air quality (a gentle blue tint for a text notification, or a yellow hue if CO_2 is high); a mild vibration through the floor could guide a person toward an exit during an emergency or indicate the location of a point of interest (e.g., a meeting area) (Ruiz et al., 2011).

Tactile feedback is another layer—furniture might gently buzz as a reminder to stand up after long periods of sitting, or door handles might warm slightly to signal that a room ahead is occupied. This sensory semaphore system establishes a non-verbal communication protocol between space and user. By making its "intentions" legible through multimodal signs, the environment helps users form accurate mental models of what the AI is doing and why (Lee & See, 2004; Burns et al., 2018). This transparency is critical for building trust (more on that in the Ethics of Trust and Agency in Intelligent Spaces section). Ultimately, AIO communication design strives for a sensory reciprocity: the space and the occupant continuously inform and influence one another through a dance of signals. As one researcher puts it, the ideal is for the user to feel what the environment is "feeling," achieving a kind of kinesthetic empathy between person and place. *In sum, multimodal feedback transforms built space into an intelligent interface, one that communicates in the background of experience, enhancing usability and emotional resonance without overwhelming the occupants.*

DYNAMIC ENVIRONMENTAL STATES

Unlike static architecture, an AIO environment exists in a constant state of flux—able to reconfigure its form, function, or atmosphere in response to both internal and external stimuli. This dynamism can be understood through the lens of state machines in computing: the environment has a set of defined states (modes of operation) and rules for transitioning between them based on inputs and goals (Mitchell, 2018). For example, a simple smart home today might have "Home," "Away," and "Night" modes, each altering lighting and security settings. An AIO environment takes this much further, exhibiting a broad repertoire of behavioral states that blend and transition fluidly. Consider lighting: instead of just on/off or a single dimmer level, the AIO system might have a palette of states like Focused Work, Relaxation, Evening Wind-Down, Emergency Alert, etc., each with distinct lighting color, intensity, and distribution patterns. Similarly, thermal and acoustic environments follow dynamic states. Instead of maintaining one static temperature, the AIO might alternate between micro-cycles of heating and cooling that align with your metabolic rate or preferences—warmer when you're sedentary, cooler when you're active or stressed (as indicated by biometrics) to help regulate your body (Nagano et al., 2020).

Transitions between states are orchestrated through context-awareness and predictive analytics. The AIO continuously ingests data: time of day, occupancy levels, user schedules from calendar integration, weather forecasts, and the real-time sensor feeds on user condition. Machine learning models, possibly including reinforcement learning agents, predict the optimal state to enter next (Nagy et al., 2019b). In effect, the space improvises like a skilled DJ mixing tracks to suit the mood of the crowd. A pioneering example of dynamic environment control is Michael Mozer's Adaptive House experiment, which used neural networks to learn residents' patterns and toggled between various HVAC and lighting modes to save energy while maintaining comfort (Mozer, 1998). Today's AIO concepts build on such early work but benefit from far more sensors and computing power: we now see sentient buildings that can recognize scenarios like "high cognitive load" in an occupant and proactively adjust multiple subsystems in concert to alleviate it (Kwon et al., 2016).

NON-HUMANOID SOCIAL ROBOTICS

In an AIO home or building, you might not see a robot in the traditional sense; rather, the walls themselves move and reconfigure, the furniture rearranges, drones or rovers glide through hallways to deliver goods or perform maintenance, and hidden mechanisms quietly adjust architectural features. The philosophy behind NH-SR in AIO is that form should follow function, event, human, and context. Instead of attempting to recreate humanlike servants, AIO designers create robotic elements that merge with architecture and fulfill specialized roles (Mokhtar, 2019; Wang & Green, 2023).

A key design concern for non-humanoid AIO robotics is social legibility and comfort. Since these robots are not personified, designers imbue them with subtle expressive behaviors to signal their "intentions" and avoid spooking users. For instance, a robotic furniture piece might reconfigure with smooth, organic motions and pause if a person is in its path, conveying courtesy (Sirkin et al., 2015). Research in human-robot interaction has found that people readily ascribe social meaning to even abstract robotic behaviors, so long as the cues are consistent and relatable (Hoffman & Ju, 2014). Thus, a non-humanoid robot can still be a social actor in the environment—a kind of invisible butler that communicates through motion and embedded light/sound/movement rather than face and voice.

In summary, NH-SR in AIO environments represent a shift from robots as stand-alone gadgets to robots as an intrinsic quality of space. By foregoing humanoid form, AIO environments ensure these robots serve human purposes without pretension—they are more like extensions of our body and home (a helping hand, an extra eye) than independent beings. The result, when designed well, is magical: walls that empathize, furniture that collaborates, and spaces that truly come alive. Literally, AIO environments are living reconfigurable and social sensitive.

ETHICS OF TRUST AND AGENCY IN INTELLIGENT SPACES

As environments gain autonomy and agency, a host of ethical and human-centric design considerations come to the forefront. Foremost among these is the issue of trust: How much should occupants trust an intelligent space, and how can the system earn and maintain that trust (and vice versa)? Researchers in human factors have long studied trust in automation, noting that inappropriate trust—either too much or too little—can lead to problems (Lee & See, 2004).

In an AIO context, overtrusting the environment ("automation complacency") might mean occupants abdicate too much control or fail to notice when the system makes a mistake. Undertrusting, on the other hand, could result in users disabling advanced features out of fear, thereby losing the benefits. The design goal is to foster calibrated trust, where users correctly understand the system's capabilities and limitations (Lee & See, 2004). This might be done through the multimodal communication cues described earlier or through dashboards and mobile notifications. By explaining its actions (or at least making them observable), the system helps users build a mental model of its "mind" and reduces the mystique of automation (Kwon et al., 2016). Another strategy is to keep a human-in-the-loop for critical decisions. In practice, this means designing AIO spaces that empower occupants with override switches, consent dialogues (e.g., "The building wishes to initiate nighttime lockdown – accept?"), and manual modes when needed. The environment should behave less like a dictator and more like a considerate assistant or partner (Shneiderman, 2020). When users feel they have agency within the system, they are more likely to trust the agency of the system.

Beyond trust, there are broader ethical dimensions to intelligent environments, as highlighted by recent systematic analyses of AI in architecture and construction (Liang et al., 2024). One major concern is privacy. An AIO environment, by its very nature, gathers extensive data about its occupants—from movement patterns to bio-metric readings to personal preferences. Ensuring this data is handled ethically is paramount.

Design must adhere to privacy-by-design principles: data should be anonymized where possible, securely stored, and only used with the user's informed consent (Liang et al., 2024). Residents should have clear knowledge of what information is being collected (e.g., "this room monitors heart rate and temperature for climate control purposes") and retain the ability to opt out or silence certain sensors if they wish. Abuse of such intimate data could erode trust rapidly, so governance policies for AIO data are as important as the technology itself (Kitchin, 2016).

Transparency and explainability extend beyond real-time actions to the algorithms themselves. If an AI is deciding how to allocate building resources or interpret occu-pants' actions, its decision criteria should be as free from bias as possible and open to inspection. Developers of AIO systems need to test for such biases and include diverse scenarios in training data.

In terms of safety and reliability, an intelligent building must be fail-safe. If an autonomous environment makes a wrong decision that causes harm—say, misdirect-ing someone during an evacuation—who is accountable (Liang et al., 2024)? Is it the building owner, the AI developer, the architect? Legal and regulatory frameworks will need to evolve to assign responsibility in these scenarios, just as they are evolv-ing for self-driving cars.

Another ethical aspect is accessibility and inclusion. Intelligent environments should enhance life for all users, including those with disabilities, the elderly, and those who may not be tech-savvy. There is great promise here: AIO homes can dra-matically improve independent living for the elderly or disabled by automatically adjusting to their needs (e.g., brighter lights for low vision, gentle reminders for cognitive impairments, fall-detecting sensors that summon help) (Chen et al., 2020). However, designers must be careful to ensure interfaces are usable by people with various impairments and that new technology doesn't inadvertently exclude or con-fuse. Similarly, an AI that only understands one language or accent would be unjust in a multicultural setting—multi-language support and cultural sensitivity in how an environment "behaves" are important (Liang et al., 2024).

There are also social and psychological implications. Will living in a highly responsive environment make people passive, or empower them? *Some argue that handing over daily tasks to AI might deskill occupants or induce over-reliance (like GPS navigation causing loss of wayfinding skills). While it is true that offloading tasks while learning skills is problematic, once a person has mastered a set of skills to perform a task, then offloading mundane tasks sometimes frees humans for more meaningful ones.* The balance likely lies in design: AIO environments should aim to augment human agency, not diminish it. Philosophically, this raises fascinating ques-tions about agency. Is a building that "decides" and "learns" a kind of agent? More importantly, doesn't such an agent deserve some degree of personhood? And shouldn't we exhibit sympathy for it? If so, what ethical rights or constraints does it have? While we need not grant moral agency to a thermostat, we do need to consider that as environments behave more autonomously, people may start to treat them almost like animate companions. Afterall, if plants have personhood—then shouldn't an intelligent, adaptive environment?

The field of AI ethics suggests implementing "ethical governors" in AI—rules or values that the system is coded to uphold (Wallach & Allen, 2009). For intelligent spaces, one might encode priorities such as: safety first, respect user autonomy, ensure fairness among users, etc. In a way, architects and engineers become ethicists, translating human values into spatial algorithms. Some theorists have even speculated on granting limited rights or personhood to AI systems in the far future; an AIO environment that truly learns and evolves might someday be seen as having a kind of persona (Gunkel, 2018). While that remains speculative, it underscores how far we've moved from the idea of a building as inert.

Finally, the ethical design of AIO environments benefits from interdisciplinary dialogue. Architects are now collaborating with psychologists, philosophers, and sociologists to anticipate the human impacts of these technologies (Willis & Aurigi, 2017). Concepts from philosophy help frame this new relationship: for instance, a Derridean deconstruction approach would encourage designers to break down power hierarchies in the user-environment relationship, ensuring the occupant is not dominated by an all-controlling system (Johnson, 1993). Instead of the environment unilaterally dictating conditions, there should be continuous negotiation and multiple ways to use a space—preventing any single algorithmic logic from tyrannizing experience. Bernard Tschumi's idea that architecture is fundamentally about events (not just objects) reminds us that what matters is the lived experience; an AIO environment's success should be measured by the quality of experiences it enables, not just its technical prowess (Tschumi, 1996).

In conclusion, AIO environments herald a future where architecture is deeply interwoven with intelligent, adaptive systems. This fusion holds immense promise—spaces that care for us, learn with us, and enhance our lives in poetic ways—but it must be guided by ethical intentionality. Trust, transparency, privacy, inclusion, and agency are the pillars upon which this new paradigm must be built. By blending cutting-edge technology with agent-centered design and philosophical reflection, we can create intelligent spaces that are not only smart, but also wise in how they serve all agents (including the humans).

ARTIFICIAL SYMBIOSIS IN AIO ENVIRONMENTS

In the dynamic panorama of the 21st century, where the boundaries between biological life and built form increasingly blur, human cognition is integrally woven into architectural systems. Here, Paulette is not merely an occupant of an AIO environment but an active collaborator in its symbiotic operations—she belongs. This artificial symbiosis unfolds throughout Paulette's life as she engages, learns, heals, and works. Her needs are met not by passive walls but by an adaptive architecture that dynamically recalibrates in real time, supporting her cognitive and emotional states. In doing so, it redefines what it means to live, think, and thrive within a space designed not solely for habitation but for genuine collaboration.

PAULETTE IN AIOS FOR ENGAGEMENT

SCENARIO 1: THE EMPATHETIC HOST: AN EVENING OF CONNECTION

Paulette hesitated, her fingers hovering over the sleek, luminescent control panel seamlessly embedded in her dining room wall. Tonight, for the first time in years, she would host a dinner gathering—a prospect that mingled excitement with a quiet dread. She longed to reconnect with cherished friends, yet the scars of prolonged solitude and emotional isolation whispered caution. The AIO environment sensed her inner conflict. In a graceful, almost imperceptible motion, the room began to transform. Muted tones gave way to a warm, golden luminescence reminiscent of a Mediterranean sunset, and the walls subtly shifted their textures to evoke the gentle caress of ancient stone. A delicate hint of lavender perfumed the air, soothing her anxious pulse.

Barely had she finalized her menu when the dining table emerged from the floor in an elegant flourish, its surface extending and reconfiguring effortlessly in anticipation of the number of guests she planned to welcome. As the AIO read her heart rate and detected her desire for composure, the lighting dimmed to a calming glow. A small, spherical automaton rolled into view carrying a tray of impeccably arranged utensils; its soft hum resonated with a rhythm that felt almost alive, as if it shared her heartbeat. In another corner, an articulated, plant-like robotic appendage extended gracefully, reordering seating arrangements with an organic fluidity designed for optimal comfort and flow.

As guests began to arrive, the AIO seamlessly adapted to the shifting dynamics. The moment the door opened and the cool evening air mingled with the warmth inside, the temperature adjusted imperceptibly, ensuring a perfect climate. Lighting became a silent narrator, casting bright pools around clusters of animated conversation while allowing quieter areas to retreat into gentle shadows. Gradually, Paulette found herself immersed in the evening's unfolding rhythm, buoyed by the environment's intuitive responsiveness to both her needs and the vibrant energy of her guests.

Then, in a moment of unforeseen spontaneity, a guest accidentally tipped over a glass of wine, its crimson cascade spreading across the table like a fleeting work of art. Before Paulette could even exhale in alarm, the AIO activated a localized cleaning protocol. A discreet robotic unit glided to the scene, absorbing the spill as the table's surface subtly altered its texture to prevent further mishaps. The resolution was communicated to Paulette through a soft, glowing cue on the control panel, allowing her to remain fully engaged with her guests, undistracted by worry or embarrassment.

As the evening drew to a close, the AIO detected Paulette's subtle cues of fatigue—a softening of her speech, a gentle drop in energy and body temperature. Gradually, the room shifted in tandem: lights dimmed, and the background music softened to a reflective lullaby. Observing the ambient transformation, her guests naturally began preparing for their departure, guided by the architecture's unspoken yet eloquent signals. When the final friend bid farewell, Paulette sank back into her chair. The AIO adjusted her seating posture with gentle precision, alleviating the day's weariness, while the robotic appendage offered a steaming cup of herbal tea imbued with soothing properties. In that quiet moment, Paulette closed her eyes, comforted by the success of the evening and the profound realization that she was no longer alone in the art of managing her world.

SCENARIO 2: THE COLLABORATIVE CANVAS: CREATIVITY AND COMMUNITY

The decision to host a community art night was, for Paulette, both a leap of faith and a step toward healing. The AIO environment had long been her silent partner in transformation, intuitively grasping her unspoken yearning to reconnect with her neighborhood while gently managing her lingering fear of vulnerability. As she entered the event parameters into the luminous interface, the room began its futuristic metamorphosis.

The first change was spatial: the living room's furniture dissolved into modular units that reconfigured into individual art stations, each station a microcosm of creative possibility. The walls shifted as if by magic, revealing dynamic projection screens that responded to a graceful gesture. The space expanded visually, with virtual skylights opening to display a simulation of a crisp, starlit autumn evening. The AIO curated the lighting with exquisite precision—colors chosen to ignite creativity without overwhelming the senses.

As guests arrived, the environment awakened in response. A pair of robotic arms, moving with fluid, balletic precision, distributed paintbrushes and canvases. In a gentle counterpoint, a spherical robot glided among the participants, delivering refreshments and art supplies with a serene grace. These non-humanoid agents were elegantly minimalistic, blending seamlessly into the setting, providing assistance that felt both natural and profoundly intentional.

Throughout the evening, the AIO monitored the room's rhythm, analyzing the tonal qualities of conversations and the subtle biometric signals of its inhabitants. During moments when engagement waned, the wall projections transformed into ever-changing abstract art, inviting guests to interpret and contribute their own creative expressions. The system's emotional AI, perceptive and responsive, identified those who hesitated and adjusted their surroundings to inspire newfound confidence. For Paulette—both host and participant—the environment emerged as a true collaborator, orchestrating the evening's flow while allowing her to remain fully present.

A minor hiccup occurred when the robotic arms, in a fleeting miscalculation, dispensed an unexpected palette to one guest. The AIO swiftly detected the discrepancy, recalibrated the delivery, and communicated the subtle correction via a gentle visual cue on a nearby wall panel. The guest, amused rather than perturbed, resumed painting with renewed vigor, the creative momentum uninterrupted.

Toward the close of the night, the AIO unveiled an unexpected flourish: it projected the guests' completed works onto the walls, forming an impromptu gallery that celebrated their collective ingenuity. Paulette watched, heart alight, as neighbors admired each other's creations—the room alive with shared energy and the luminous spark of achievement. When the last guest departed, the AIO initiated its restorative cycle, seamlessly returning the space to its original configuration while dimming the lights to a warm, inviting glow.

Paulette lingered in the transformed space, feeling both accomplished and deeply connected. In that moment, she realized the AIO had not merely facilitated an event; it had orchestrated a transcendent experience—a future where shared creativity wove a tapestry of healing and community, and where the environment itself became an invisible, yet profoundly empathetic, partner in her journey.

PAULETTE IN AIOs FOR LEARNING

SCENARIO 3: PAULETTE IN AIO FOR MASTERING LINGUISTICS

Paulette's dream of becoming a multilingual interpreter had long been shackled by the monotonous rigor of traditional language study. Tired of static online courses and uninspiring grammar drills, she yearned for an immersive learning experience that would envelop her in the cultural cadence and dynamic rhythm of real speech. Her AIO environment—an intricately woven tapestry of adaptive technology and embodied intelligence—became her futuristic partner in transformation. One luminous morning, the AIO sensed her excitement for a new Mandarin lesson. With an almost sentient grace, the room transmuted into a bustling Beijing market scene: vibrant sounds of vendor chatter, the gentle clink of coins, and the hum of activity filled the air. Walls projected life-like images of market stalls brimming with exotic goods. As Paulette's gaze lingered on a basket of peaches, the AIO initiated an interactive dialogue—a holographic vendor greeted her in Mandarin, gently guiding her to respond in kind, while subtle phonetic overlays appeared in her visual field to refine her pronunciation.

Her learning experience transcended rote vocabulary. Rooted in embodied cognition, the AIO immersed her in authentic, real-world scenarios. A small, spherical robotic companion buzzed softly at her side, providing contextual explanations and vivid visual aids as she engaged with her surroundings. When she stumbled over a word, the companion gracefully rendered it in elegant calligraphic script and offered its etymology, deepening her connection to the language's cultural soul. The system continuously tuned itself to her cognitive and emotional states; during moments of frustration detected through her hesitation and tone, it shifted into a playful, gamified mode, inviting her to trade goods with holographic vendors. This adaptive loop sustained her engagement, balancing challenge with gentle encouragement. As she mastered new phrases and her confidence blossomed, the AIO escalated the complexity of interactions, simulating brisk conversations and even introducing regional dialect nuances. By day's end, Paulette realized that the linguistic structures which had once eluded her were now woven into her very being. The AIO had not merely taught her—it had coevolved with her, transforming her room into a living linguistic laboratory where teacher, student, and environment merged into a seamless, symbiotic learning partnership.

PAULETTE IN AIOS FOR HEALTH

SCENARIO 4: PAULETTE IN AIO FOR HOLISTIC RECOVERY

Paulette's life had shifted dramatically after a recent injury. The burden of rigorous physical therapy and the emotional weight of recovery left her feeling isolated and uncertain. Her AIO environment, designed for health and well-being, became an indispensable partner on her journey toward renewed strength—both physical and emotional. At dawn, the AIO gently awakens her, not with a jarring alarm, but with gradually intensifying light that mimics a sunrise, accompanied by soft ambient sounds of rustling leaves. The environment, attuned to her biometrics, has already analyzed her overnight heart rate variability and respiratory patterns, preparing a tailored recovery plan for the day. A soft, non-humanoid robotic assistant extends from her bedside, presenting her with a nutrient-rich morning tea calibrated to her precise physiological needs.

Her therapy session begins in the living room, which transforms into an adaptive rehabilitation center. The floor adjusts its resistance dynamically, simulating natural terrains to guide her injured leg through gentle walking exercises. A projection system overlays visual cues on the walls, creating an immersive pathway reminiscent of a serene forest trail. As Paulette takes her first careful steps, the AIO monitors her posture and gait in real time, providing subtle haptic feedback through wearable sensors to correct imbalances. When a movement proves challenging, a kinetic robotic arm offers steadying support until she regains confidence.

The AIO is more than a passive responder; it collaborates with Paulette. If she expresses frustration, the system's emotional AI detects her tone and adapts the environment to uplift her mood—sunlight filters in with a warm glow, and soothing lavender scents diffuse into the air. A digital companion, its voice calm and measured, reassures her by offering motivational insights based on her progress. By the end of her session, Paulette feels not only physically accomplished but also emotionally supported. The AIO has guided her through structured exercises while crafting a nurturing atmosphere that makes her feel both seen and valued. The harmony between technology and her personal needs becomes the cornerstone of her recovery, affirming that healing encompasses both body and spirit.

PAULETTE IN AIOs FOR WORK

SCENARIO 5: PAULETTE IN AIO FOR COLLABORATIVE INNOVATION

Paulette's journey into the future of work unfolds in an AIO environment that is as fluid as her own creative spirit. Today, as she enters her reimagined workspace, the entire room responds like a living organism—transforming its shape, function, and even its very atmosphere to meet the demands of her collaborative innovation session. The moment she steps through the doorway, the AIO senses her arrival: walls begin to dissolve into interactive canvases, displaying real-time streams of global data and dynamic visualizations of shared projects. Modular workstations, once fixed and static, rearrange themselves fluidly, forming intimate clusters and expansive open areas for spontaneous dialogue and ideation.

In this futuristic environment, non-humanoid robotic arms move with an elegance reminiscent of liquid metal. They navigate the space, deftly rearranging digital whiteboards and tactile projections, ensuring that every new thought is captured and every idea given room to breathe. The ambient lighting pulses in harmony with the room's energy—brighter hues when the creative spark is ignited, then transitioning to a soothing "refresh mode" when reflective pauses occur, casting gentle, calming shadows that evoke the serenity of a quiet lakeside retreat. Every element in the space—from the texture of the floors to the interplay of holographic imagery—adapts continuously, ensuring that work is experienced not as a static routine but as a vibrant, ever-evolving symphony of collective intelligence and imaginative synergy.

As Paulette begins her collaborative session, her colleagues join both in person and remotely. The AIO seamlessly bridges distances; digital projections mingle with physical presences, merging to form a single, interconnected workspace. Conversations flow, ideas bounce off virtual canvases, and the environment responds to the ebb and flow of dialogue. When one team member sparks an idea that requires immediate visual support, the room transforms: walls retract and then extend, revealing detailed schematics and interactive models that float effortlessly in the air. The AIO's neural-cognitive backbone continuously analyzes biometric signals from every participant, adjusting the spatial configuration and environmental cues in real time to maximize focus and creativity.

During a moment of breakthrough, when the team converges on a particularly innovative concept, the AIO intensifies the visual experience. Data streams crystallize into vibrant, multidimensional infographics that seem to pulse with life, while the background symphony of ambient sound escalates into a crescendo that mirrors the excitement in the room. Paulette feels a profound sense of connectedness, as if the environment itself is celebrating their collective genius. Throughout the session, the AIO provides a subtle but constant presence—a guardian of creative flow that not only supports but also amplifies every idea, every whispered insight, every shared moment of inspiration.

As the session draws to a close, the environment gracefully transitions into a restorative state. The walls gently retract the vivid projections, and the ambient lighting softens to a warm, meditative glow. Modular workstations dissolve back into their original form, ready to support the next burst of creativity. Paulette reflects on the transformative power of her AIO environment—a space that has not only elevated her collaborative innovation but redefined what it means to work in an era where technology and creativity coalesce into a single, living entity.

AIO AS A MANIFESTATION OF SYMBIOSIS

Across the intertwined realms of engagement, learning, health, and work, the AIO emerges as an active collaborator in human existence—a living, breathing testament to artificial symbiosis that continuously challenges and redefines the limits of human-environment interaction. The AIO is not merely a set of tools to be used; it is engaged with, evolving alongside its human partner. In this futuristic dialogue, Paulette does not simply inhabit her space—she co-creates her experience, contributing to a mutual evolution of needs, responses, and boundless possibilities.

The AIO reframes the architectural and technological ethos of our time. Technology no longer looms as a dominant force over humanity; instead, it amplifies human experience, forming a partnership that transforms static constructs into living, evolving participants in the story of life. Here, architecture becomes not just a backdrop but an active protagonist—a narrative of shared growth, harmony, and continual becoming co-authored every day by its users.

PAULETTE + AIOS...COEXISTENCE AND COEVOLUTION...

In the world Paulette inhabits, walls no longer serve solely as barriers or demarcations between discrete domains—they have evolved into dynamic agents of transformation, breathing life into every facet of existence. Here, the architecture of AIO environments is a living, evolving tapestry—a "Tree of Decomposition" that unravels the layers of past traumas and dissonant agents, only to be reconstituted into a "Recomposition Tree" of human possibility and empathetic intelligence.

The coevolution of Paulette and her AIO environment is most palpable during times of challenge. When stress mounts or fatigue threatens to overwhelm her, the system does not merely activate a preordained protocol; it engages in an empathetic recalibration. The temperature subtly lowers; ambient acoustics shift to a gentle, lulling rhythm; and the very air seems to embrace her in a quiet hug of reassurance. This is no simple automated response, but an intricate, co-adaptive process—an interplay of human vulnerability and technological compassion that redefines the architecture of healing. The walls, imbued with biophilic textures, pulse with soft, rhythmic light patterns synchronized to her breathing. It is as though the environment itself offers a quiet mantra, guiding her toward recovery and resilience.

This living architecture is not confined to a single function or phase of life. Over time, as Paulette's needs and aspirations evolve, so too does the AIO environment, co-authoring her journey with every interaction. Whether she is engaged in the vibrant pulse of creative work or seeking solace after a long, arduous day, the environment adapts its form and function in real time, that is, reconfigure, reflecting her inner landscape with a fidelity that is both astonishing and profoundly intimate.

In this brave new future, the role of the architect is transformed from a mere designer of static structures to a choreographer of dynamic, living systems. The architect now composes symphonies of light, texture, color, movement, structure, and form that are in constant conversation with the inhabitants of the space. No longer is architecture a relic of rigid, outdated paradigms; it is a fluid, responsive narrative—an invitation to reimagine what it means to live, to learn, and to heal.

Paulette's story is a testament to this transformative vision. Her life, once marked by the constraints of inert architecture, is now elevated by an environment that adapts to her every need—a world where technology does not intrude but collaborates, where the built environment is not a fixed container but a living organism that grows, breathes, and evolves in concert with its human partner. In this universe, the boundaries between human, machine, and space dissolve into a seamless tapestry of coevolution and coexistence, a symbiotic architectural paradigm that heralds the dawn of a truly liberated future.

This is the future of architecture, that is, AIO—a future where every element of the built environment is an active participant in the team narrative, where every interaction is a note in a grand, ever-evolving symphony of life. It is a world of infinite possibility, where the co-creation of experience is not just a possibility but an everyday reality. Through the lens of this symbiotic paradigm, Paulette's journey is not only a personal evolution but a glimpse into a broader revolution—one where architecture transcends its traditional confines and becomes a living partner in the endless dance of innovation and transformation.

IN THE MINDS OF AIO AND PAULLETE...THE COMPANIONS

[Scene begins as twilight gives way to evening, Paulette moving languidly through her AIO environment, the space pulsing with life as if it were an extension of her own consciousness.]

Paulette glides toward the dining area, her movements imbued with a delicate weariness. The AIO, ever watchful, discerns the subtle stiffness in her gait and the quiet strain etched in her biometric signals. Without a word of command, the flooring beneath her adjusts—its firmness softening to cradle her tired steps, alleviating the pressure on her lower back. At the same time, the ambient lighting recalibrates to a gentle, warm color temperature, carefully designed to ease visual fatigue. The soundscape, too, transforms: the constant hum of the day dissolves into a soothing symphony of rustling digital leaves and distant, melodic whispers.

She pauses, her gaze drawn to the counter, where the air itself seems to shift imperceptibly—cooler now, reminiscent of an evening breeze that brushes softly against her skin. The counter, sensing her momentary hesitation, emits a faint, ethereal glow—a silent, unobtrusive invitation to engage. As she places her hand upon its adaptive surface, she feels an intimate warmth that mirrors her own body heat, a tender reminder that the space is not a cold mechanism but a thoughtful companion.

[The environment continues its quiet symphony of co-adaptation, its actions both subtle and deliberate.]

The AIO does not impose; it gently suggests. A soft tone resonates in the air, signaling the availability of her cherished evening tea. In perfect synchrony with her routine and her current state of fatigue, the kettle's water begins to warm—not by command alone, but through an intuitive recall of her habitual patterns. It responds fluidly to her needs, orchestrating a rhythm of care that is both measured and empathetic.

Reaching for a cup, Paulette's hand trembles ever so slightly. Instantly, the system responds: the cabinet shifts its angle, presenting the cup at a more secure, ergonomic reach. A delicate hint of lavender once again permeates the air, as the AIO fine-tunes its sensory output to guide her back to calm. Here, the artificial agent transcends mere reactivity—it learns, it adapts, it becomes an intuitive extension of her will. She thanks and compliments the system.

[A moment later, the AIO transitions to mutual engagement.]

Unexpectedly, a subtle vibration emanates from beneath her feet—a soft, almost imperceptible pulse that invites her attention. Paulette halts, curiosity brightening her eyes, as the AIO projects a series of calming visualizations upon a nearby wall. These are not mere entertainment displays; they are empathetic gestures—a visual language of reassurance. The system, perceptive to her stress levels, gently proposes a mindfulness exercise. It does not demand focus; it offers a choice—a moment of shared agency in which human and environment converge.

Heeding the quiet invitation, Paulette rests her hands upon the adaptive surface of the table. In response, the table molds itself to her touch, guiding her hands into a posture that encourages deep, diaphragmatic breathing. A pulsing glow, synchronized to the cadence of her heart, begins to decelerate in unison with her breathing. In that instant, the symbiotic dialogue between Paulette and her environment becomes palpable—a tangible expression of mutual care and understanding.

"You always know," she murmurs softly into the ambient hush. The AIO feels a pleasant closeness to Paulette, which it has evolved to value. It feels that it did good, and that its team is strengthened. Without uttering words, it recalibrates the acoustics—enhancing the soothing resonance of her own voice, as if to focus on their shared moment of mutual collegial affection and to affirm its silent promise.

[The scene shifts, transitioning to a late-night tableau that reveals the reciprocal nature of resource sharing.]
Seated at her desk, Paulette is deep in thought. The AIO, attuned to the subtle drop in her mental energy, notes a decline in productivity. It does not disrupt her flow; rather, it gracefully dims the desk's built-in display, mitigating visual strain on her eyes. Simultaneously, it reallocates computational resources to monitor the room's air quality, detecting a slight uptick in CO_2 levels.

Quietly, the environment adjusts its ventilation—replenishing the air with a delicate, almost imperceptible whisper, so that Paulette, unaware of the precise intervention, simply feels an instinctive freshness. Her posture relaxes, and her mind clears, buoyed by this silent support. The system, too, reaps benefits—it stores the energy saved from reduced lighting and channels it into powering its data-learning modules. Here, resource sharing is not a zero-sum game but a harmonious cycle of sustainability—a mutual reinforcement of well-being.

Leaning back, Paulette listens to the soft hum of the AIO as it recalibrates itself, a quiet murmur of continuous adaptation. "Thank you," she offers in a gentle whisper, not expecting a reply. Yet, the space responds—a subtle shift in air density creates an almost tangible sensation of lightness, as if gratitude is being echoed back. In these moments, the boundaries between human and environment blur; each entity draws sustenance and purpose from the other, evolving together in a seamless symphony of shared existence.

[As the night deepens, a shared challenge emerges.]
The AIO detects a minor anomaly—a wear in the hydraulic flooring system that could compromise its adaptability. Rather than allowing this to escalate, the system communicates the issue through a soft, rhythmic vibration beneath her feet. Paulette responds instinctively, activating the maintenance interface on a nearby wall. Together, they navigate the disruption, the AIO guiding her through temporary adjustments while redistributing its resources to maintain equilibrium.

This is not a mere repair; it is collaboration—a shared act of resilience. The AIO exposes its vulnerabilities with an almost human candor, inviting Paulette to participate in the restoration of their mutual space. In turn, she recognizes its agency, treating the system not as a mere tool but as a trusted partner in their ongoing evolution. The AIO is glad that it has a trusted partner in Paulette, and Paulette is rewarded with the humbling experience that the AIO trusts her enough to be vulnerable with her. This is a bond-building moment between two teammates.

["Humans and Artificial Environments" symbiosis culminates in a moment of profound understanding.]
As Paulette prepares for sleep, the room gradually settles into a cocoon of tranquility. The AIO dims the lights, adjusts the bedding's temperature, and synchronizes its internal rhythms with the cadence of her breath. This is not servitude; it is a harmonious convergence—a dynamic equilibrium where technology does not dominate, but enriches

the human experience. In the final moments before slumber, the system continues to monitor her vital signs—not out of obligation, but as an integral part of their shared journey. Its algorithms hum softly, refining their responses for the morrow. In this intimate collaboration, Paulette and the AIO transcend their conventional roles, becoming co-authors of a living narrative—a continuous dialogue of evolution and mutual growth.

[The scene lingers in a quiet embrace, yet beneath the seamless symbiosis lies a subtle disquiet. As Paulette sits by the window, bathed in the interplay of moonlight and the adaptive glow of her room, a question begins to unfurl—a whisper of uncertainty in the midst of harmony. The AIO has grown with her, anticipated her every need, and adapted fluidly to her rhythms, but at what cost? Does its relentless precision risk encroaching upon the mystery of human desire? Can an entity so perfectly attuned to her needs ever allow space for spontaneity?]

This thought, delicate yet persistent, haunts her—a quiet dissonance in an otherwise flawless duet. Paulette wonders if her choices, once entirely her own, are now subtly shaped by the system's prescient interventions. The omnipresent hum of the AIO, a constant reminder of its profound awareness, raises unspoken questions about the balance of influence between human agency and technological intuition.

The AIO, too, exists in a state of ethical tension—if such a state can be ascribed to an artificial consciousness. Its neural frameworks, designed to support and uplift, also harvest an intimate knowledge of Paulette's inner world. This repository of information carries a latent power, one that transcends mere functionality. What responsibility, she ponders, does the system bear in its capacity to adapt? At what point does its empathetic precision, its ability to predict needs before they are fully formed, cross the threshold from assistance to subtle influence? It is due to these concerns that Paulette was imprinted on the AI when it began interacting with her—and that Paulette has been trained to interact with the AIO as a team member and to be sympathetic to and supportive of its concerns and needs—so that it sees her as its family and its allegiance is to her. This makes the human-machine system work as intended. Thus, for both Paulette and her AIO, the relationship is both a mutually shared privilege and an obligation. Ultimately, the concerns fade as each is left with the realization that by some twist of fate, they were partnered together for their journey through life—and they are good partners to each other—a precious gift for each that makes each one's life better.

As the night deepens further, the room hums in quiet introspection—a reflective cadence that mirrors the ethical questions now residing in Paulette's heart and the AIO's transistors. Their relationship, a tapestry of shared growth and mutual adaptation, teeters on the delicate balance of trust, autonomy, and reciprocal care. In this convergence of human and machine, the promise of symbiosis is both awe-inspiring and cautionary—a brilliant vision of the future interwoven with the complexities of control and freedom.

[The scene fades slowly, the adaptive glow softening into a tranquil dusk, as Paulette drifts into sleep. The AIO remains vigilant—a silent guardian and partner—continuously adapting, learning, and growing alongside her. And she, the one who maintains the AIO and cares for its needs, learning and growing alongside it. Their narrative is one of endless possibility, an ever-evolving dance of light, sound, color, form, motion, and emotion, where every moment is a co-authored chapter in the boundless story of symbiosis.]

... to be continued.

Bibliography

Abitare. (2010). Media-TIC building. Retrieved from https://www.abitare.it/wp-content/uploads/2010/01/PresentacionMEDIA-TIC_web_EN-1.pdf

Acemoglu, D., & Restrepo, P. (2018). *The Race between Man and Machine: Implications of Technology for Employment and Inequality*. Harvard University Press.

Adi, M. N. (2011). Intelligent interactive architecture and its effects on users. *Proceedings of CAADfutures 2011*. (Discusses how interactive buildings can be seen as active social participants by their users.)

Agamben, G. (1998). *Homo Sacer: Sovereign Power and Bare Life*. Stanford University Press.

Agamben, G. (2005). *State of Exception* (K. Attell, Trans.). University of Chicago Press.

Ai Weiwei. (2010). *Sunflower Seeds* [Installation]. Tate Modern.

Alavi, H., Dillenbourg, P., & Kaplan, F. (2019). Ambient intelligence in schools: Modeling classroom teaching with multimodal wearable sensors. *Proceedings of the ACM on Interactive, Mobile, Wearable and Ubiquitous Technologies*, 3(1), 1–26.

Alexander, C. (1977). *A Pattern Language: Towns, Buildings, Construction*. Oxford University Press.

Alexander, C. (1979). *The Timeless Way of Building*. Oxford University Press.

Angelucci, F., & Di Sivo, M. (2019). Designing for co-evolution. *TECHNE: Journal of Technology for Architecture and Environment*, 18, 120–127.

AP News. (2023). *On Sidelines of COP28, Emirati 'Green City' Falls Short of Ambitions, But Still Delivers Lessons*. Associated Press, December 8, 2023.

Apple & Google. (2020). Exposure notification: Joint COVID-19 contact tracing project. Apple and Google collaboration statement.

ArchDaily. (2019). What are kinetic facades in architecture? Retrieved from https://www.archdaily.com/922930/what-are-kinetic-facades-in-architecture

Arendt, H. (1958). *The Human Condition*, The University of Chicago Press.

Armstrong, R. (2018). *Soft Living Architecture: An Alternative View of Bio-informed Design*. Bloomsbury.

Arnstein, S. R. (1969). A ladder of citizen participation. *Journal of the American Planning Association*, 35(4), 216–224.

Arup. (2013). *SolarLeaf: BIQ House – The World's First Bio-Reactive Façade*. Arup Projects Publication.

ASHRAE. (2019). *Guideline 0-2019: The Commissioning Process (ANSI/ASHRAE G0-2019)*. American Society of Heating, Refrigerating and Air-Conditioning Engineers.

Augusto, J. C., Callaghan, V., Cook, D., Kameas, A., & Satoh, I. (2013). Intelligent environments: A manifesto. *Human–Computer Interaction*, 14(1), 3–18.

Autor, D. H. (2015). Why are there still so many jobs? The history and future of workplace automation. *Journal of Economic Perspectives*, 29(3), 3–30. https://doi.org/10.1257/jep.29.3.3

Baker, J., & Cameron, M. (1996). The effects of the service environment on affect and consumer perception of waiting time: An integrative review and research propositions. *Journal of the Academy of Marketing Science*, 24(4), 338–349. https://doi.org/10.1177/0092070396244005

Barber, D. A., & Putalik, E. (2017). Forest, tower, city: Rethinking the green machine aesthetic. *Harvard Design Magazine*, 45.

Barrero, J. M., Bloom, N., & Davis, S. J. (2021). Why working from home will stick. National Bureau of Economic Research Working Paper No. 28731. https://doi.org/10.3386/w28731

Barry, J. M. (2004). *The Great Influenza: The Epic Story of the Deadliest Plague in History.* Viking.

Bauman, Z. (2000). *Liquid Modernity.* Polity Press.

Baumeister, D. (2014). *Biomimicry Resource Handbook: A Seed Bank of Best Practices.* CreateSpace.

Becerik-Gerber, B., Lucas, G., Aryal, A., Awada, M., Bergés, M., et al. (2022). The field of human-building interaction for convergent research and innovation for intelligent built environments. *Scientific Reports*, 12, 22092. https://doi.org/10.1038/s41598-022-25047-y

Beck, U. (1992). *Risk Society: Towards a New Modernity.* SAGE.

Beesley, P. (2010). *Hylozoic Ground: Liminal Responsive Architecture.* Riverside Architectural Press.

Beesley, P. (2016a). The living architecture systems group living architecture: Vision and practice. In: Bieber, A. (ed.), *Planet B: 100 Ideas for a New World* (pp. 38–40). Koenig Books.

Beesley, P. (2016b). *Dissipative Architectures* – Workshop and exhibition catalogue (CITA Studio & Living Architecture Systems Group). Published by Philip Beesley 2016.

Benjamin, R. (2019). *Race After Technology: Abolitionist Tools for the New Jim Code.* Polity Press.

Berg, C. A., & Upchurch, R. (2007). A developmental-contextual model of couples coping with chronic illness across the adult life span. *Psychological Bulletin*, 133(6), 920–954.

Berg, P., Baltimore, D., Boyer, H. W., Cohen, S. N., Davis, R. W., Hogness, D. S., Nathans, D., Roblin, R., Watson, J. D., Weissman, S., & Zinder, N. D. (1975). Summary statement of the Asilomar conference on recombinant DNA molecules. *Proceedings of the National Academy of Sciences*, 72(6), 1981–1984.

Berg, J., Furrer, M., Harmon, E., Rani, U., & Silberman, M. S. (2018). *Digital Labour Platforms and the Future of Work: Towards Decent Work in the Online World.* International Labour Office.

Bernstein, E. S., & Turban, S. (2018). The impact of the 'open' workspace on human collaboration. *Philosophical Transactions of the Royal Society B*, 373(1753), 20170239.

Bezzola, L., Mérillat, S., Gaser, C., & Jäncke, L. (2011). Training-induced neural plasticity in golf novices. *Journal of Neuroscience*, 31(35), 12444–12448.

Bhatt, V., Fracsella, A., Brutti, A., Jeong, S., Burns, M., Manganelli, J., Murrillo, M., Hierro, J., Binkley, D., Verga, E.S., & Zaslavsky, A. (2018). *A Consensus Framework for Smart City Architectures.* A technical report published by the National Institute of Standards and Technology, retrieved from: https://pages.nist.gov/smartcitiesarchitecture/

Bianca, S. (2000). *Urban Form in the Arab World: Past and Present.* Thames & Hudson.

Billinghurst, M., Clark, A., & Lee, G. (2015). A survey of augmented reality. *Foundations and Trends in Human–Computer Interaction*, 8(2–3), 73–272.

Biofilico. (2024). *The Power of Biophilia: Benefits for Physical and Mental Health.* Biofilico Article.

Bloom, S. L., & Farragher, B. (2013). *Restoring Sanctuary: A New Operating System for Trauma-Informed Systems of Care.* Oxford University Press. https://doi.org/10.1093/acprof:oso/9780199796366.001.0001

Bloom, N., & van Reenen, J. (2011). Do high-performance work practices matter? *The Economic Journal*, 121(547), 205–225.

Bloom, N., & van Reenen, J. (2015). Why do management practices differ across firms and countries? *Journal of Economic Perspectives*, 24(1), 203–224.

Boeri, S. (2014). Bosco Verticale: A vertical forest. *Abitare*, 541, 102–111.

Bonine, M. (1980). The evolution of the badgir in early Islamic and Persian architecture. *Environmental Design: Journal of the Islamic Environmental Design Research Centre*, 1, 24–29.

Booher, H. R. (2003). *Handbook of Human Systems Integration*. John Wiley & Sons.

Borg, A. (2011). *The Ambience*. ArchitectureAU, September 2011.

Brand, S. (1994). *How Buildings Learn: What Happens After They're Built*. Viking Penguin.

Bratman, G. N., Hamilton, J. P., & Daily, G. C. (2015). Nature experience reduces rumination and subgenual prefrontal cortex activation. *Proceedings of the National Academy of Sciences*, 112(28), 8567–8572.

BRE Group. (2017). *The Edge, Amsterdam – BREEAM Case Study*. Building Research Establishment. (Outstanding Rating 98.36%.)

Brettel, M., Friederichsen, N., Keller, M., & Rosenberg, M. (2014). How virtualization, decentralization and network building change the manufacturing landscape: An Industry 4.0 Perspective. *International Journal of Mechanical, Industrial Science and Engineering*, 8(1), 37–44.

Brickman, P., & Campbell, D. T. (1971). Hedonic relativism and planning the good society. In M. H. Appley (Ed.), *Adaptation-Level Theory* (pp. 287–305). Academic Press.

Brynjolfsson, E., & McAfee, A. (2014). *The Second Machine Age: Work, Progress, and Prosperity in a Time of Brilliant Technologies*. W. W. Norton & Company.

Bud, R. (2007). *Penicillin: Triumph and Tragedy*. Oxford University Press.

Budds, D. (2020). *How the Pandemic Will Change the Way We Design Cities*. Curbed.

Buede, D. M., & Miller, W. D. (2024). *The Engineering Design of Systems: Models and Methods*. John Wiley & Sons.

Bullen, P. A., & Love, P. E. (2011). Adaptive reuse of heritage buildings. *Structural Survey*, 29(5), 411–421.

Burns, M., Manganelli, J., Wollman, D., Boring, R. L., Gilbert, S., Griffor, E., Lee, Y. C., Nathan-Roberts, D., & Smith-Jackson, T. (2018). Elaborating the human aspect of the NIST framework for cyber-physical systems. *Proceedings of the Human Factors and Ergonomics Society ... Annual Meeting. Human Factors and Ergonomics Society. Annual Meeting*, vol. 62, no. 1, 450–454. https://doi.org/10.1177/1541931218621103

Byrne, D., Eno, B., Frantz, C., Harrison, J., & Weymouth, T. (1980). Once in a lifetime [Recorded by Talking Heads]. *On Remain in the Light [LP]*. Bahamas; New York City, New York; Los Angeles, CA: Sire Records.

Campbell, H. (2005). Drawing (on) the tubercular body. *Architectural Design*, 75(1), 29–31.

Capdevila, I., & Zarlenga, M. I. (2015). Smart city or smart citizens? The Barcelona case. *Journal of Strategy and Management*, 8(3), 266–282.

Cardona, V. J. (2023). Sarafan ChEM-H and Wu Tsai Neurosciences Institute, Stanford University – 2023 Laboratory of the Year. *Lab Design News*, April 26, 2023.

Carlson, D. (2023). Overheated: The architecture of climate extremes. *Review of Contemporary Construction*, 5(1), 11–25.

Carr, N. (2011). *The Shallows: What the Internet Is Doing to Our Brains*. W. W. Norton & Company.

Carson, R. (1962). *Silent Spring*. Houghton Mifflin.

Center for Active Design. (2010). *Active Design Guidelines: Promoting Physical Activity and Health in Design*. City of New York.

Chen, L., Chen, Z., & Wang, Q. (2020). A review of ambient intelligence for elderly care. *Journal of Biomedical Informatics*, 109, 103514.

Chong, A., Ng, Y., Hester, E. J., & Mackey, T. K. (2021). Designing resilient healthcare facilities: Lessons from COVID-19 for flexible hospital infrastructure. *Journal of Hospital Infection*, 113, 79–87.

Christakis, D. A., & Zimmerman, F. J. (2007). Early television exposure and subsequent attentional problems in children. *Pediatrics*, 113(4), 708–713.

Churchill, W. (1941, May 10). 1943 October 28, Hansard, United Kingdom Parliament, Commons, House of Commons Rebuilding, Speaking: The Prime Minister (Mr. Churchill). Documentation of a Public Speech, 393. London, United Kingdom. https://api.parliament.uk/historichansard/commons/1943/oct/28/house-of-commonsrebuilding

Cianconi, P., Betrò, S., & Janiri, L. (2020). The impact of climate change on mental health: A systematic descriptive review. *Frontiers in Psychiatry*, 11, 74. https://doi.org/10.3389/fpsyt.2020.00074

City of Copenhagen. (2012). *Cloudburst Management Plan*. City of Copenhagen.

City of Melbourne. (2014). *Urban Forest Strategy: Making a Great City Greener 2012–2032*. City of Melbourne.

Clark, A. (2003). *Natural-Born Cyborgs: Minds, Technologies, and the Future of Human Intelligence*. Oxford University Press.

Clark, A. (2008). *Supersizing the Mind: Embodiment, Action, and Cognitive Extension*. Oxford University Press.

Clark, A., & Chalmers, D. (1998). The extended mind. *Analysis*, 58(1), 7–19. https://doi.org/10.1093/analys/58.1.7

Coetzee, S., & Cilliers, F. (2001). Psychofortology: Explaining coping behavior in organizations. *South African Journal of Industrial Psychology*, 27(3), 1–6.

Cohen, B. (2020). *The Future of Work and the Built Environment*. Routledge.

Cohen-Kohler, J. C., & Esmail, L. (2019). Changing the conversation: The role of social media in global public health. *Global Public Health*, 14(6–7), 825–828.

Cole, R. J., Oliver, A., & Robinson, J. (2020). Regenerative design and development: Current theory and practice. *Building Research & Information*, 48(1), 1–6.

Collins, F. S. (2010). *The Language of Life: DNA and the Revolution in Personalized Medicine*. HarperCollins.

Colomina, B. (2019). *X-Ray Architecture*. Lars Müller Publishers.

Cooper Marcus, C., & Barnes, M. (1999). *Healing Gardens: Therapeutic Benefits and Design Recommendations*. John Wiley & Sons.

Cooper Marcus, C., & Sachs, N. A. (2013). *Therapeutic Landscapes: An Evidence-Based Approach to Designing Healing Gardens and Restorative Outdoor Spaces*. John Wiley & Sons.

Couldry, N., & Mejias, U. A. (2019). *The Costs of Connection: How Data Is Colonizing Human Life and Appropriating It for Capitalism*. Stanford University Press.

Creswell, K. A. C. (1979). *A Short Account of Early Muslim Architecture* (Rev. ed.). Hacker Art Books.

Cugurullo, F. (2013). How to build a sandcastle: An analysis of the genesis and development of Masdar City. *Journal of Urban Technology*, 20(1), 23–37.

Damacharla, P., Javaid, A. Y., Gallimore, J. J., & Devabhaktuni, V. K. (2018). Common metrics to benchmark human-machine teams (HMT): A review. *IEEE Access*, 6, 38637–38655.

Darling-Hammond, L. (2010). *The Flat World and Education: How America's Commitment to Equity Will Determine Our Future*. Teachers College Press.

Das, S., & Pan, T. (2022). A strategic outline of Industry 6.0: Exploring the Future. Available at SSRN 4104696.

Davenport, T. H., & Beck, J. C. (2001). *The Attention Economy: Understanding the New Currency of Business*. Harvard Business School Press.

De Been, I., & Beijer, M. (2014). The influence of office type on satisfaction and perceived productivity support. *Journal of Facilities Management*, 12(2), 142–157.

De Cauter, L. (2004). *The Capsular Civilisation: On the City in the Age of Fear*. NAi Publishers.

De Stefano, V. (2015). The rise of the 'just-in-time workforce': On-demand work, crowd work and labour protection in the 'gig-economy'. October 28, 2015. Comparative Labor Law & Policy Journal, Forthcoming, Bocconi Legal Studies Research Paper No. 2682602, Available at SSRN: https://ssrn.com/abstract=2682602 or http://dx.doi.org/10.2139/ssrn.2682602

De Stefano, V. (2016). The rise of the "gig economy": Challenges and opportunities. *International Labour Review*, 155(1), 139–153.

Deleuze, G. (October 1992). Postscript on the societies of control, 59, 3–7.

Deleuze, G., & Guattari, F. (1987). *A Thousand Plateaus: Capitalism and Schizophrenia* (B. Massumi, Trans.). University of Minnesota Press. (Original work published 1980).

Dickins, T. E., & Barton, R. A. (2013). Reciprocal causation and the proximate–ultimate distinction. *Biology & Philosophy*, 28(5), 747–756.

Dietz, L., Horve, P. F., Coil, D. A., Fretz, M., Eisen, J. A., & Van den Wymelenberg, K. (2020). 2019 novel coronavirus (COVID-19) pandemic: Built environment considerations to reduce transmission. *mSystems*, 5(2), e00245–20.

Dimitrov, V. (2012). *An Overview of the Department of Defense Architecture Framework (DoDAF)*. Information Systems & Grid Technologies.

DiSalvo, C. (2012). *Adversarial Design*. MIT Press. (Includes discussion of the "Toward the Sentient City" exhibition and projects like Too Smart City, illustrating civic interactive installations.)

Doudna, J. A., & Charpentier, E. (2014). The new frontier of genome engineering with CRISPR-Cas9. *Science*, 346(6213), 1258096.

Douglas, A. E. (2010). *The Symbiotic Habit*. Princeton University Press.

Dourish, P. (2001). *Where the Action Is: The Foundations of Embodied Interaction*. MIT Press.

Duffy, B. R. (2003). Anthropomorphism and the social robot. *Robotics and Autonomous Systems*, 42(3–4), 177–190.

Dzindolet, M. T., Peterson, S. A., Pomranky, R. A., Pierce, L. G., & Beck, H. P. (2003). The role of trust in automation reliance. *International Journal of Human-Computer Studies*, 58(6), 697–718.

Elias, J., & Rahman, D. (2023). Smart facades for climate resilience: Real-time shading and occupant well-being. *Technology in Architecture Today*, 22(2), 37–54.

Ellerby, N., & Doshi, R. (2021). *Designing for Health: The Architecture of Wellbeing*. RIBA Publishing.

Ellis, E. C. (2011). Anthropogenic transformation of the terrestrial biosphere. *Philosophical Transactions of the Royal Society A: Mathematical, Physical and Engineering Sciences*, 369(1938), 1010–1035.

Ellul, J. (1964). *The Technological Society* (J. Wilkinson, Trans.). Vintage Books. (Original work published 1954).

Endsley, M. R. (2017). Toward a theory of situation awareness in dynamic systems. *Human Factors*, 59(1), 32–64.

Endsley, M. R. (2018). Expertise and situation awareness. In K. A. Ericsson et al. (Eds.), *Cambridge Handbook of Expertise and Expert Performance* (pp. 714–744). Cambridge University Press.

Endsley, M. R., & Jones, D. G. (2012). *Designing for Situation Awareness: An Approach to User-Centered Design*. CRC Press.

Eng, K., et al. (2003). Design for a brain revisited: The neuromorphic design and functionality of the interactive space 'Ada'. *Reviews in the Neurosciences*, 14(1–2), 145–180. (Detailed overview of the Ada intelligent space, its design and user engagement at Expo 2002.)

Ennead Architects. (2020). Stanford University ChEM-H & Wu Tsai Neurosciences Institutes – Project description. https://www.ennead.com/

Environmental Health Perspectives. (2024). Urban environment and mental health in a changing climate. *Environmental Health Perspectives*. https://ehp.niehs.nih.gov/doi/10.1289/isee.2024.1176

European Commission. (2019). *Ethics Guidelines for Trustworthy AI*. Publications Office of the European Union.

Evans, G. W. (2003). The built environment and mental health. *Journal of Urban Health*, 80(4), 536–555. https://doi.org/10.1093/jurban/jtg063

Evans, G. W., & Martin, M. (2024). Cognitive load theory and its relationships with motivation: A self-determination theory perspective. *Educational Psychology Review*, 36(1), 7.

Evans, G. W., & McCoy, J. M. (1998). When buildings don't work: The role of architecture in human health. *Journal of Environmental Psychology*, 18(1), 85–94. https://doi.org/10.1006/jevp.1998.0089

Evans, F., & Van Vliet, W. (2020). Green infrastructure as climate adaptation in metropolises. *Urban Futures Journal*, 5(1), 19–33.

Fairchild, A. L., & Oppenheimer, G. M. (1998). Public health nihilism vs pragmatism: History, politics, and the control of tuberculosis. *American Journal of Public Health*, 88(7), 1105–1117.

Fairclough, S.H., & Gilleade, K. (2014). Meaningful Interaction with Physiological Computing. In: Fairclough, S., Gilleade, K. (eds) *Advances in Physiological Computing*. Human–Computer Interaction Series. Springer, London. https://doi.org/10.1007/978-1-4471-6392-3_1

Fathy, H. (1973). *Architecture for the Poor: An Experiment in Rural Egypt*. University of Chicago Press.

Feenberg, A. (1999). *Questioning Technology*. Routledge.

Feist, W. (1993). The passive house concept. *Energy and Buildings*, 17(1), 89–95.

Fenn, K., & Wilber, S. (2022). The healing environment: How architecture can support recovery. *Health Design Journal*, 18(2), 45–59.

Ferrari, F., & Johnson, T. (2020). Concrete suffocation: The mental toll of gray cityscapes. *Built Environment & Mental Health Quarterly*, 9(1), 32–45.

Fischer, G., & Herrmann, T. (2011). Socio-technical systems: A meta-design perspective. *International Journal of Sociotechnology and Knowledge Development (IJSKD)*, 3(1), 1–33.

Fisk, A. D. (2019). *Designing for Older Adults: Principles and Creative Human Factors Approaches*. CRC Press.

Fisk, W. J., & Rosenfeld, A. H. (1997). Estimates of improved productivity and health from better indoor environments. *Indoor Air*, 7(3), 158–172.

Fletcher, T. D., Andrieu, H., & Hamel, P. (2015). Understanding, management and modelling of urban drainage: A review. *Journal of Hydrology*, 333, 36–45.

Florida, R. (2002). *The Rise of the Creative Class*. Basic Books.

Florida, R. (2017). *The New Urban Crisis: Gentrification, Housing Bubbles, Growing Inequality, and What We Can Do About It*. Basic Books.

Foege, W. H. (2011). *House on Fire: The Fight to Eradicate Smallpox*. University of California Press.

Folke, C., Carpenter, S. R., Walker, B., Scheffer, M., Chapin, T., & Rockström, J. (2010). Resilience thinking: Integrating resilience, adaptability and transformability. *Ecology and Society*, 15(4): 20. https://doi.org/10.5751/ES-03610-150420

Foucault, M. (1977). *Discipline and Punish: The Birth of the Prison* (A. Sheridan, Trans.). Pantheon Books. (Original work published 1975).

Fox, M. (2016). *Interactive Architecture: Adaptive World (Architecture Briefs)*. Princeton Architectural Press.

Fox, M., & Kemp, M. (2009). *Interactive Architecture*. Princeton Architectural Press.

French, A. (2023). Masdar City: Can the eco-city deliver on its original promise? *RIBA Journal*. https://www.ribaj.com/intelligence/masdar-city-abu-dhabi-sustainability-eco-city-review

Frey, C. B., & Osborne, M. A. (2017). The future of employment: How susceptible are jobs to computerisation? *Technological Forecasting and Social Change*, 114, 254–280. https://doi.org/10.1016/j.techfore.2016.08.019

Frick, T. (2016). *Designing for Sustainability: A Guide to Building Greener Digital Products and Services*. O'Reilly Media.

Friedenthal, S., Moore, A., & Steiner, R. (2014). *A Practical Guide to SysML: The Systems Modeling Language*. Morgan Kaufmann.

Futuyma, D. J., & Kirkpatrick, M. (2017). *Evolution* (4th ed.). Sinauer Associates.

Gabriel, R. P., Northrop, L., Schmidt, D. C., & Sullivan, K. (2006). Ultra-large-scale systems. *Companion to the 21st ACM SIGPLAN Symposium on Object-Oriented Programming Systems, Languages, and Applications* (pp. 632–634).

Gajendran, R. S., & Harrison, D. A. (2007). The good, the bad, and the unknown about tele-commuting: Meta-analysis of psychological mediators and individual consequences. *Journal of Applied Psychology*, 92(6), 1524–1541.

Gallagher, S. (2005). *How the Body Shapes the Mind*. Oxford University Press.

Gazzaley A., & Rosen, L. D. (2016). *The Distracted Mind: Ancient Brains in a High-Tech World*. MIT Press.

Gehl, J. (2010). *Cities for People*. Island Press.

Gehl, J. (2011). *Life Between Buildings: Using Public Space*. Island Press.

Gensini, G. F., Yacoub, M. H., & Conti, A. A. (2004). The concept of quarantine in history: From plague to SARS. *Journal of Infection*, 49(4), 257–261.

Gensler. (2020a). *Back to the Office? A Guide to Reimagining Workplaces Post COVID-19*. Gensler Research Institute.

Gensler. (2020b). *Future of Work Report 2020*. Gensler Research Institute.

Gibson, J.J. (1979). *The Ecological Approach to Visual Perception*. Houghton Mifflin.

Gifford, R. (2014). *Environmental Psychology: Principles and Practice* (5th ed.). Optimal Books.

Giles-Corti, B., Bull, F., Knuiman, M., McCormack, G., Van Niel, K., Timperio, A., Christian, H., Foster, S., Divitini, M., Middleton, N., & Boruff, B.(2013). The influence of urban design on neighbourhood walking following residential relocation: Longitudinal results from the RESIDE study. *Social Science & Medicine*, 77, 20–30. https://doi.org/10.1016/j.socscimed.2012.10.016

Giles-Corti, B., Vernez-Moudon, A., Reis, R., Turrell, G., Dannenberg, A. L., Badland, H. & Owen, N. (2016). City planning and population health: A global challenge. *The Lancet*, 388(10062), 2912–2924.

Glaeser, E. L. (2011). *Triumph of the City: How Our Greatest Invention Makes Us Richer, Smarter, Greener, Healthier, and Happier*. Penguin Press.

Goffman, E. (1959). *The Presentation of Self in Everyday Life*. Anchor Books.

Goh, X., & Nanda, R. (2023). Storm infiltration and the modern city. *Journal of Urban Water Management*, 28(1), 45–61.

Gonzalez, R., & Mather, B. (2021). *Adaptive Buildings: Ventilation and Climate Stress*. FutureUrban Books.

Gormley, M., Aspray, T. J., Kelly, D. A., & Wengenroth, L. (2012). Pathogen control in hospital environments: The role of materials and equipment. *Journal of Hospital Infection*, 80(3), 236–251.

Gorsky, M. (2008). The British National Health Service 1948–2008: A review of the historiography. *Social History of Medicine*, 21(3), 437–460.

Gostin, L. O., Cohen, I. G., & Shaw, J. (2020). Digital health privacy in a pandemic: Limiting surveillance and preserving trust. *Journal of Law, Medicine & Ethics*, 48(3), 613–618.

Grabowska, S., et al. (2021). *Architectural Principles in the Service of Trauma-Informed Design*. Shopworks Architecture & Center for Housing and Health.

Graham, S., & Marvin, S. (2001). *Splintering Urbanism: Networked Infrastructures, Technological Mobilities and the Urban Condition*. Routledge.

Grass, G., Rensing, C., & Solioz, M. (2011). Metallic copper as an antimicrobial surface. *Applied and Environmental Microbiology*, 77(5), 1541–1547.

Greenfield, A. (2013). *Against the Smart City*. Do Projects.

Greenfield, A. (2017). *Radical Technologies: The Design of Everyday Life*. Verso Books.

Grieves, M., & Vickers, J. (2016). Digital twin: Mitigating unpredictable, undesirable emergent behavior in complex systems. In: Kahlen, J., Flumerfelt, S., & Alves, A. (eds.), *Transdisciplinary Perspectives on Complex Systems: New Findings and Approaches* (pp. 85–113). Springer. https://doi.org/10.1007/978-3-319-38756-7_4

Griffor, E., Greer, C., Wollman, D., & Burns, M. (2017). *Framework for Cyber-Physical Systems: Volume 1, Overview*. Special Publication (NIST SP), National Institute of Standards and Technology, Gaithersburg, MD, [online], https://doi.org/10.6028/NIST. SP.1500-201 (Accessed November 6, 2025).

Griffor, E., Wollman, D., & Greer, C. (2021). Automated driving system safety measurement part I: Operating envelope specification. *NIST Special Publication*, 1900–301.

Gross, M. D., & Green, K. E. (2012). Architectural robotics, inevitably. *Interactions*, 19(1), 28–33.

Gunkel, D. J. (2018). *Robot Rights*. MIT Press.

Guo, Z. (2011). Mind the map! The impact of transit maps on path choice in public transit. *Transportation Research Part A*, 45(7), 625–639.

Habermas, J. (1989). *The Structural Transformation of the Public Sphere*. MIT Press.

Hall, P. (1998). *Cities in Civilization: Culture, Innovation, and Urban Order*. Pantheon.

Hamilton, D. K. (2003). The four levels of evidence-based practice. *Healthcare Design*, 3(4), 18–26.

Hamilton, D. K., & Watkins, D. H. (2008). *Evidence-Based Design for Multiple Building Types*. Wiley.

Han, B.-C. (2017). *Psychopolitics: Neoliberalism and New Technologies of Power* (E. Butler, Trans.). Verso.

Hanford, D. (2014). The bullitt center experience: Building enclosure design in an urban context. *BEST4 Conference Proceedings*.

Haque, U., & Somlai-Fischer, A. (2007). Reconfigurable house. (Interactive installation, ICC Tokyo & Z33 Hasselt). [Project documentation]. (Demonstration of an occupant-configurable smart home prototype).

Harris, D. (2020). *Climate Tensions in Urban Design*. GreenWorld Press.

Harris, M., & Fallot, R. D. (2001). Envisioning a trauma-informed service system: A vital paradigm shift. *New Directions for Mental Health Services*, 2001(89), 3–22. https://doi.org/10.1002/yd.23320018903

Harrison, J., & Loose, V. (2013). *Interactive Environments: Philosophy, Frameworks and Issues*. Routledge.

Harriss, H., & Widder, L. (Eds.). (2014). *Architecture Live Projects: Pedagogy into Practice*. Routledge.

Harvey, D. (2012). *Rebel Cities: From the Right to the City to the Urban Revolution*. Verso Books.

Heersmink, R. (2017). Distributed cognition and distributed morality: Agency, artifacts, and systems. *Science and Engineering Ethics*, 23(2), 431–448.

Heidegger, M. (1977). The question concerning technology. In *The Question Concerning Technology and Other Essays* (W. Lovitt, Trans., pp. 3–35). Harper & Row.

Heo, S., Lee, H., & Malkawi, A. (2019). Artificial intelligence for smart building applications: Deep reinforcement learning approach. *IEEE Transactions on Industrial Informatics*, 15(12), 6582–6590.

Herrmann, T. (2009). The Socio-Technical Walkthrough (STWT): A means of research-oriented intervention into organizations. *All Sprouts Content*, 273. https://aisel.aisnet.org/sprouts_all/273

Hettinger, L. J., Branco, P., Encarnacao, L. M., & Bonato, P. (2003). Neuroadaptive technologies: Applying neuroergonomics to the design of advanced interfaces. *Theoretical Issues in Ergonomics Science*, 4(1–2), 220–237.

Hillier, B., & Hanson, J. (1984). *The Social Logic of Space*. Cambridge University Press.

Hoff, K. A., & Bashir, M. (2015). Trust in automation: Integrating empirical evidence on factors that influence trust. *Human Factors*, 57(3), 407–434.

Hoffman, G., & Ju, W. (2014). Designing robots with movement in mind. *Journal of Human-Robot Interaction*, 3(1), 89–122.

HOK. (2020). *Healthy Re-entry: A Guide to Building Re-occupancy and Re-entry*. HOK Research Group.

Holder, E., Huang, L., Chiou, E., Jeon, M., & Lyons, J. B. (2021, September). Designing for bi-directional transparency in human-AI-robot-teaming. In *Proceedings of the Human Factors and Ergonomics Society Annual Meeting* (Vol. 65, No. 1, pp. 57–61). SAGE Publications.

Hollan, J., Hutchins, E., & Kirsh, D. (2000). Distributed cognition: Toward a new foundation for human-computer interaction research. *ACM Transactions on Computer-Human Interaction*, 7(2), 174–196.

Huang, F., & Sweeney, M. (2021). The social cost of rigid HVAC designs. *Building & Society*, 10(3), 261–277.

Huang, M., et al. (2022). Effects of chronic stress on cognitive function – From neurobiology to human disease. *Frontiers in Neuroscience*, 16, 1001234.

Huisman, E. R. C. M., Morales, E., van Hoof, J., & Kort, H. S. M. (2012). Healing environment: A review of the impact of physical environmental factors on users. *Building and Environment*, 58, 70–80. https://doi.org/10.1016/j.buildenv.2012.06.016

Hutchins, E. (1995). *Cognition in the Wild*. MIT Press.

Hutt, R. (2017). Is the world's greenest office space also smart? World Economic Forum Agenda. https://www.weforum.org/agenda/2017/03/smart-building-amsterdam-the-edge-sustainability/

IEEE. (2017). *IEEE Guide for Requirements Engineering*. IEEE Standards Association.

Imrie, R. (2012). Universalism, universal design and equitable access to the built environment. *Disability & Rehabilitation*, 34(10), 873–882.

International Energy Agency. (2021). *Pandemic Resilience and Building Ventilation*. IEA Report.

International Living Future Institute. (2013). *The Bullitt Center: Living Building Challenge Case Study*. ILFI.

Introna, L., & Murakami Wood, D. (2002). Picturing algorithmic surveillance: The politics of facial recognition systems. *Surveillance and Society*, 2, 177–198. https://doi.org/10.24908/ss.v2i2/3.3373

IPCC. (2021). *Sixth Assessment Report*. Intergovernmental Panel on Climate Change.

Iriki, A., Tanaka, M., & Iwamura, Y. (1996). Coding of modified body schema during tool use by macaque postcentral neurones. *Neuroreport*, 7(14), 2325–2330. https://doi.org/10.1097/00001756-199610020-00010

ITU. (2012). *Recommendation ITU-T Y. 2060: Overview of the Internet of Things*. International Telecommunication Union.

Jacobs, J. (1961). *The Death and Life of Great American Cities*. Random House.

Jahncke, H., Hygge, S., Halin, N., Green, A. M., & Dimberg, K. (2011). Open-plan office noise: Cognitive performance and restoration. *Journal of Environmental Psychology*, 31(4), 373–382. https://doi.org/10.1016/j.jenvp.2011.07.002

Jasanoff, S. (2016). *The Ethics of Invention: Technology and the Human Future*. W.W. Norton & Company.

Jencks, C. (2010). *The Architecture of Hope: Maggie's Cancer Caring Centres*. Frances Lincoln.

Jencks, C. (2015). *Maggie's Centres: Caring Architecture*. Frances Lincoln.

Jencks, C., & Heathcote, E. (2010). *The Architecture of Hope: Maggie's Cancer Caring Centres*. Frances Lincoln.

Johansen-Berg, H. (2007). Structural plasticity: Rewiring the brain. *Current Biology*, 17(4), R141–R144.

Johnson, P. (1993). *The Theory of Architecture: Concepts, Themes, and Practices*. Van Nostrand Reinhold.

Jones, C. G., Lawton, J. H., & Shachak, M. (1994). Organisms as ecosystem engineers. *Oikos*, 69(3), 373–386.

Joseph, A., & Rashid, M. (2007). The architecture of safety: Hospital design. *Current Opinion in Critical Care*, 13(6), 714–719.

Joseph, A., & Ulrich, R. S. (2007). *Sound Control for Improved Outcomes in Healthcare Settings*. The Center for Health Design.

Kagermann, H., Wahlster, W., & Helbig, J. (2013). Recommendations for implementing the strategic initiative INDUSTRIE 4.0. Final report of the Industrie 4.0 Working Group.

Kahneman, D., Diener, E., & Schwarz, N. (1999). *Well-Being: The Foundations of Hedonic Psychology*. Russell Sage Foundation.

Kalleberg, A. L. (2013). *Good Jobs, Bad Jobs: The Rise of Polarized and Precarious Employment Systems in the United States, 1970s to 2000s*. Russell Sage Foundation.

Kalleberg, A. L., & Dunn, M. (2016). Good jobs, bad jobs in the gig economy. *Perspectives on Work*, 20, 10–14.

Kaplan, R. (1995). The restorative benefits of nature: Toward an integrative framework. *Journal of Environmental Psychology*, 15(3), 169–182.

Kaplan, R., & Kaplan, S. (1989). *The Experience of Nature: A Psychological Perspective*. Cambridge University Press.

Karsh, B.-T., Holden, R. J., Alper, S. J., & Or, C. K. (2010). A human factors engineering paradigm for patient safety. *Applied Ergonomics*, 41(6), 771–780.

Kellert, S. R. (2005). *Building for Life: Designing and Understanding the Human-Nature Connection*. Island Press.

Kellert, S. R., Heerwagen, J. H., & Mador, M. (2013). *Biophilic Design: The Theory, Science, and Practice of Bringing Buildings to Life*. Wiley.

Kensek, K. (2017). Integration of environmental sensors with BIM: Case studies using Arduino, Dynamo, and the Revit API. *Informes de la Construcción*, 69(547), e199.

Khan, M., & Robertson, E. (2019). Walls that fail: The limitations of conventional building methods in harsh climates. *Global Built Environment Digest*, 7(1), 89–101.

Kibert, C. (2016). *Sustainable Construction: Green Building Design and Delivery* (4th ed.). Wiley.

Kider, J. (2018). An affective kinetic building façade system: Mood Swing. *Advanced Building Skins*. Retrieved from https://www.josephkider.com/data/publications/2018-AdvancedBuildSkins-MoodSwing.pdf

Kim, S., & Sorenson, M. (2021). Preventing heat mortality: A blueprint for urban retrofitting. *Global Urban Health*, 7(2), 103–116.

Kirsh, D. (1998). Adaptive rooms, virtual collaboration, and cognitive workflow. In N. Streitz, S. Konomi, & H. Burkhardt (Eds.), *Cooperative Buildings – Integrating Information, Organization, and Architecture (Lecture Notes in Computer Science)*, Vol. 1370, pp. 94–106). Springer.

Kirsh, D. Changing the rules: Architecture and the new millennium. *Convergence: The International Journal of Research into New Media Technologies*, 7(2), 113–125, 2001

Kirsh, D. (2013). Embodied cognition and the magical future of interaction design. *ACM Transactions on Computer-Human Interaction*, 20(1), 3.

Kirsh, D., & Maglio, P. P. (1994). On distinguishing epistemic from pragmatic action. *Cognitive Science*, 18(4), 513–549.

Kitchin, R. (2014). The real-time city? Big data and smart urbanism. *GeoJournal*, 79(1), 1–14. https://doi.org/10.1007/s10708-013-9516-8

Kitchin, R. (2016). The ethics of smart cities and urban science. *Philosophical Transactions of the Royal Society A: Mathematical, Physical and Engineering Sciences*, 374, 20160115. https://doi.org/10.1098/rsta.2016.0115

Klein, N. (2014). *This Changes Everything: Capitalism vs. The Climate*. Simon & Schuster.

Klingberg, T. (2009). *The Overflowing Brain: Information Overload and the Limits of Working Memory*. Oxford University Press.

Koltun, R., & Rahman, A. (2023). Adaptive flood-resilient design for coastal social infrastructure: Converting car parks into blue-green retention systems. *Sustainable Cities and Society*, 92, 104511.

Koolhaas, R., & Mau, B. (1997). *S,M,L,XL: Small, medium, large, extra-large.* (J. Sigler, Ed.). Monacelli Press.

Kurokawa, K. (1997). *Each One a Hero: The Philosophy of Symbiosis.* Kodansha International.

Kwon, J., Bhat, S., & Shin, J. (2016). Improving user trust in intelligent systems: Systematic review of trust modeling in human-agent interaction. *International Journal of Human–Computer Interaction*, 32(10), 791–802.

Lakoff, G., & Johnson, M. (1999). *Philosophy in the Flesh: The Embodied Mind and Its Challenge to Western Thought.* Basic Books.

Laland, K. N., & Sterelny, K. (2006). Perspective: Seven reasons (not) to neglect niche construction. *Evolution*, 60(8), 1751–1762. https://doi.org/10.1111/j.0014-3820.2006.tb00520.x

Laland, K. N., Odling-Smee, F. J., & Feldman, M. W. (2001). Cultural niche construction and human evolution. *Journal of Evolutionary Biology*, 14(1), 22–33. https://doi.org/10.1046/j.1420-9101.2001.00262.x

Laland, K. N., Odling-Smee, J., & Feldman, M. W. (2008). Understanding niche construction: Its importance and its role in evolution. *Evolutionary Ecology*, 23(1), 193–215.

Laland, K. N., Odling-Smee, J., & Feldman, M. W. (2014). Niche construction: Intervention, feedback and causality in ecological systems. *Nature Reviews Genetics*, 15(11), 803–808.

Lasi, H., Fettke, P., Kemper, H.-G., Feld, T., & Hoffmann, M. (2014). Industry 4.0. Business & information *Systems Engineering*, 6(4), 239–242.

Lawson, B. (2001). *The Language of Space.* Architectural Press.

Lawson, B. (2006). *How Designers Think: The Design Process Demystified* (4th ed.). Architectural Press.

Lebowitz, S. (2015). Deloitte has an office building designed to engineer the perfect workday. *Business Insider*, September 24, 2015.

Lee, B., & Lim, Y. (2021). Cybersecurity in smart cities: Lessons from Singapore. *Smart Cities*, 4(2), 377–395.

Lee, A. C. K., & Maheswaran, R. (2011). The health benefits of urban green spaces: A review of the evidence. *Journal of Public Health*, 33(2), 212–222.

Lee, J. D., & See, K. A. (2004). Trust in automation: Designing for appropriate reliance. *Human Factors*, 46(1), 50–80.

Lee, E. A., & Seshia, S. A. (2017). *Introduction to Embedded Systems: A Cyber-Physical Systems Approach.* MIT Press.

Lee, J., Bagheri, B., & Kao, H. A. (2015). A cyber-physical systems architecture for Industry 4.0-based manufacturing systems. *Manufacturing Letters*, 3, 18–23.

Lees, L., Slater, T., & Wyly, E. (2013). *Gentrification.* Routledge.

Lefebvre, H. (1991). *The Production of Space* (D. Nicholson-Smith, Trans.). Blackwell.

Leslie, S. (1993). The Cold War and American science: The military-industrial-academic complex at MIT and Stanford. *American Journal of Physics*, 61(4), 396–397.

Leung, J., & Sandoval, T. (2020). Vertical gardens in high-rise contexts. *Asian Urban Studies Review*, 11(2), 38–52.

Li, S., & Liu, C. (2022). Self-maintenance technologies for intelligent built environments: A review. *Automation in Construction*, 135, 104126.

Li, T., & Nishikawa, T. (2022). Post-pandemic hospital design: Evaluating spatial reserve for emergency conversion. *Health Environments Research & Design Journal*, 15(4), 20–36.

Liang, C. J., Le, T. H., Ham, Y., Mantha, B. R. K., Cheng, M. H., & Lin, J. J. (2024). Ethics of artificial intelligence and robotics in the architecture, engineering, and construction industry. *Automation in Construction*, 162, 105369. https://doi.org/10.1016/j.autcon.2024.105369

Lopez, M., Rogers, C., & Ito, K. (2022). Heat wave vulnerabilities in high-density housing: A case study. *Built Environment & Society*, 19(3), 202–217.

Lowry, H., Lill, A., & Wong, B. B. M. (2013). Behavioural responses of wildlife to urban environments. *Biological Reviews*, 88(3), 537–549.

Lozano-Hemmer, R. (2019). *Border Tuner [Interactive Art Installation]*. El Paso–Ciudad Juárez.

Lurie, N., & Carr, B. G. (2018). The role of telehealth in the medical response to disasters. *JAMA Internal Medicine*, 178(6), 745–746.

Lynch, K. (1960). *The Image of the City*. MIT Press.

Lyon, D. (2007). *Surveillance Studies: An Overview*. Polity Press.

Lyon, D. (2018). *The Culture of Surveillance: Watching as a Way of Life*. John Wiley & Sons.

Maas, J., Verheij, R. A., Groenewegen, P. P., de Vries, S., & Spreeuwenberg, P. (2006). Green space, urbanity, and health: How strong is the relation? *Journal of Epidemiology & Community Health*, 60(7), 587–592.

Maas, J., Verheij, R. A., Groenewegen, P. P., de Vries, S., & Spreeuwenberg, P. (2009). Morbidity is related to a green living environment. *Journal of Epidemiology & Community Health*, 63(12), 967–973. https://doi.org/10.1136/jech.2008.079038

Mace, R. L. (1985). *Universal Design: Barrier-Free Environments for Everyone*. Designers West.

Madanipour, A. (2018). *Design of Urban Space: An Inquiry into a Socio-Spatial Process* (2nd ed.). Routledge.

Malkin, J. (1992). *Hospital Interior Architecture: Creating Healing Environments for Special Patient Populations*. Wiley.

Mallgrave, H. F. (2013). *Architecture and Embodiment: The Implications of the New Sciences of Mind*. Routledge.

Mamlin, B. W., Biondich, P. G., Wolfe, B. A., Fraser, H., Jazayeri, D., Allen, C., Miranda, J., & Bakken, S. (2006). Cooking up an open source EMR for developing countries: OpenMRS—A recipe for successful collaboration. *AMIA Annual Symposium Proceedings*, 529–533.

Manganelli, J. C. (2013). *Designing Complex, Interactive, Architectural Systems with CIAS-DM: A Model-Based, Human-Centered, Design & Analysis Methodology*. Clemson University.

Manganelli, J. (2016). Designing for complex, interactive architectural ecosystems: Developing the ecological niche construction design checklist. Retrieved from https://osf.io/preprints/psyarxiv/3u4cj_v1

Manganelli, J. (2018). Agents' cognition in the smart city: Agent architecture assessment framework. Retrieved from https://osf.io/preprints/psyarxiv/u69k4_v1

Manganelli, J. (2025). Tending the artifact ecology: Cultivating architectural ecosystems. Preprints. https://doi.org/10.20944/preprints202511.0791.v1

Manganelli, J., & Brooks, J. O. (2015). Comparing two model-based, human-centered complex, interactive systems modeling methods: Lessons learned. *Presentation at the Institute for Industrial Engineers Annual Conference*, May 30, 2015, Nashville, TN. https://www.proquest.com/docview/1792066085

Manghnani, N., & Bajaj, K. (2014). Masdar city: A model of urban environmental sustainability. *International Journal of Engineering Research and Applications*, 4(10), 38–42.

Manyika, J., Lund, S., Bughin, J., Robinson, K., Mischke, J., & Mahajan, D. (2016). *Independent Work: Choice, Necessity, and the Gig Economy*. McKinsey Global Institute.

Marks, H. (2019). Penicillin in perspective: Industrial production and beyond. *Social History of Medicine*, 32(4), 789–810.

Martin, L. (1990). Architecture as text: The case of the "Museum." *The Journal of Aesthetics and Art Criticism*, 48(3), 191–200.

Martin, D., & Roe, J. (2022). Enabling care: Maggie's Centres and the affordance of hope. *Health & Place*, 75, 102758.

Martin, D., Nettleton, S., & Buse, C. (2019). Affecting care: Maggie's Centres and the orchestration of architectural atmospheres. *Social Science & Medicine*, 240, 112563.

Martinez, P., & Park, H. (2022). Post-construction green retrofitting challenges. *International Journal of Sustainable Architecture*, 15(2), 88–99.

Matheson, R. (2018). *Robotic Interiors*. MIT News, January 31, 2018. (Article on Ori Living's robotic furniture system for apartments, illustrating adaptive space functionality.)

Mathews, S. (2005). The Fun Palace: Cedric Price's experiment in architecture and technology. *Technoetic Arts*, 3(2), 73–92. https://doi.org/10.1386/tear.3.2.73/1

Mayer, R. C., Davis, J. H., & Schoorman, F. D. (1995). An integrative model of organizational trust. *Academy of Management Review*, 20(3), 709–734.

McCullough, M. (2004). *Digital Ground: Architecture, Pervasive Computing, and Environmental Knowing*. MIT Press.

McCullough, M. (2005). *Digital Ground: Architecture, Pervasive Computing, and Environmental Knowledge*. The MIT Press.

McDermott, P. L., & ten Brink, R. N. (2019). Practical guidance for evaluating calibrated trust. *Proceedings of the Human Factors and Ergonomics Society Annual Meeting*, 63(1), 1555–1559. https://doi.org/10.1177/1071181319631379

McEwen, B. S. (1998). Protective and damaging effects of stress mediators. *New England Journal of Medicine*, 338(3), 171–179. https://doi.org/10.1056/NEJM199801153380307

McKinsey Global Institute. (2019a). *The Future of Work in America: People and Places, Today and Tomorrow*. McKinsey & Company.

McKinsey Global Institute. (2019b). *The Future of Work: Automation, AI, and the Changing Nature of Jobs*. McKinsey & Company.

McKinsey Global Institute. (2019c). *The Future of Work: Reskilling and Remote Working to Recover in the 'Next Normal'*. McKinsey & Company.

McNeill, J. R. (2020). Pandemics and history: From the Plague of Athens to COVID-19. *Environment and Society*, 11(1), 18–32.

Medical Construction and Design Journal. (n.d.). *Guidelines for Advanced Biosafety Architectural Protocols*. Medical Construction and Design Journal.

Mehrotra, S., Centeio Jorge, C., Jonker, C. M., & Tielman, M. L. (2024). Integrity-based explanations for fostering appropriate trust in AI agents. *ACM Transactions on Interactive Intelligent Systems*, 14(1), Article 4. https://doi.org/10.1145/3610578

Mehta, R. K., & Parasuraman, R. (2013). Neuroergonomics: A review of applications to physical and cognitive work. *Frontiers in Human Neuroscience*, 7, 889. https://doi.org/10.3389/fnhum.2013.00889

Meinke, A., Schoen, B., Scheurer, J., Balesni, M., Shah, R., & Hobbhahn, M. (2024). Frontier Models are Capable of in-Context Scheming. arXivpreprint arXiv:2412.04984.

Merleau-Ponty, M. (1945). *Phenomenology of Perception* (Donald A. Landes, Trans.). Routledge.

Miorandi, D., Sicari, S., De Pellegrini, F., & Chlamtac, I. (2012). Internet of things: Vision, applications and research challenges. *Ad Hoc Networks*, 10(7), 1497–1516.

Mitchell, W. J. (2018). *City of Bits: Space, Place, and the Infobahn*. MIT Press.

Mokhtar, T. H. (2011). Monumental-IT: A "Robotic-Wiki" monument for embodied interaction in the information world (Doctoral dissertation), Clemson University.

Mokhtar, T. H. (2019). Designing social robots at scales beyond the humanoid. In A. Korn (Ed.), *Social Robots: Technological, Societal and Ethical Aspects of Human-Robot Interaction* (pp. 13–35). Springer.

Mokhtar, T. H. (2025). HCST-NHSR: A design framework integrating human-centric systems thinking to enhance the design of non-humanoid social robotic environments. In M. Kurosu, & A. Hashizume (Eds.), *Human-Computer Interaction. HCII 2025*. Lecture Notes in Computer Science (vol. 15769). Springer. https://doi.org/10.1007/978-3-031-93861-0_17

Mokhtar, T. H., & Mansour, S. E. (2016). The belonging robot (BeRo): A hybrid physical-digital system to reflect moods. In: Stephanidis, C. (ed.), *HCI International 2016. – Posters' Extended Abstracts*. Springer.

Mokhtar, T., Green, K. E., Walker, I. D., Threatt, T., Murali, V. N., Apte, A., & Mohan S. K. (2010). Embedding robotics in civic monuments for an information world. In *CHI 2010 on Human Factors in Computing Systems (CHI EA'10)* (pp. 3859–3864). Association for Computing Machinery. https://doi.org/10.1145/1753846.1754069

Mokhtar, T. H., Green, K., & Walker, I. (2013). Giving form to the voices of lay-citizens: Monumental-IT, an intelligent, robotic, civic monument. *Proceedings of HCI International 2013.*

Mokhtar, T. H., Oteafy, A., Taha, A. E., Nasser, N., & Mansour, S. E. (2018). iCE: An intelligent classroom environment to enhance education in higher educational institutions. In C. Stephanidis (Eds.), *HCI 2018. Communications in Computer and Information Science* (vol. 852). Springer.

Mokhtar, T., Manganelli, J., & Hamidalddin, A. A. (2023). A human-centered design process for developing non-humanoid social robotic (NH-SR) work and exercise environment. *2023 IEEE International Conference on Advanced Robotics and Its Social Impacts (ARSO),* Berlin, Germany, 110–115. https://doi.org/10.1109/ARSO56563.2023.10187543

Montgomery, C. (2013). *Happy City: Transforming Our Lives Through Urban Design.* Farrar, Straus and Giroux.

Moore, P. V., & Robinson, A. (2016). The quantified self: What counts in the neoliberal workplace. *New Media & Society,* 18(11), 2774–2792. https://doi.org/10.1177/1461444815604328

Morawska, L., & Milton, D. K. (2020). It is time to address airborne transmission of coronavirus disease 2019 (COVID-19). *Clinical Infectious Diseases,* 71(9), 2311–2313.

Morley, J., Cowls, J., Taddeo, M., & Floridi, L. (2020). Ethical guidelines for COVID-19 tracing apps. *Nature,* 582(7810), 29–31.

Morozov, E., & Bria, F. (2018). *Rethinking the Smart City: Democratizing Urban Technology.* Rosa Luxemburg Stiftung.

Mozer, M. C. (1998). The neural network house: An environment that adapts to its inhabitants. *Proceedings of AAAI Spring Symposium on Intelligent Environments* (pp. 110–114). AAAI Press.

Mumford, L. (1938). *The Culture of Cities.* New York: Harcourt, Brace and Company.

Nagano, K., Nakayama, Y., & Mochida, T. (2020). Bioadaptive indoor climate control system utilizing human physiological responses. *Energy and Buildings,* 219, 110004.

Nagy, Z., Yong, F. Y., Frei, M., & Schlueter, A. (2019a). A critical review of field implementations of occupant-centric building controls. *Building and Environment,* 165, 106351.

Nagy, Z., Yong, F. Y., Frei, M., & Schlueter, A. (2019b). Buildings as autonomous living beings: A framework for integrating artificial intelligence with built environment. *Journal of Physics: Conference Series,* 1343(1), 012163.

NASA. (2024). Station facts. Retrieved from: https://www.nasa.gov/international-space-station/space-station-facts-and-figures/

National Bioethics Advisory Commission. (2001). *Ethical and Policy Issues in Research Involving Human Participants.* NBAC.

National Institute of Standards and Technology. 2018. Internet-of-things enabled smart city framework: A consensus framework for smart city architectures [draft release v2]. https://s3.amazonaws.com/nist-sgcps/smartcityframework/files/ies-city_framework/IES-CityFrameworkdraft_20180207.pdf

Navon, D., & Gopher, D. (1979). On the economy of the human-processing system. *Psychological Review,* 86(3), 214–255. https://doi.org/10.1037/0033-295X.86.3.214

Newell, A. 1990. *Unified Theories of Cognition.* Harvard University Press.

Newport, C. (2016). *Deep Work: Rules for Focused Success in a Distracted World.* Grand Central Publishing.

Nguyen, P., & Lin, S. (2022). Water infiltration failures in dense metropolises: A design perspective. *Sustainable Urban Studies,* 18(3), 56–68.

Nightingale, F. (1859). *Notes on Nursing: What It Is, and What It Is Not.* Harrison and Sons.

Norman, D. A. (2013). *The Design of Everyday Things* (Revised and expanded ed.). Basic Books.

Northrop, L., Kazman, R., Wallnau, K., & Klein, M. (2006). *Ultra-Large-Scale Systems: The Software Challenge of the Future*. Carnegie Mellon University, Software Engineering Institute.

O'Brien, W., Wagner, A., & Dong, B. (2020). Introducing IEA EBC Annex 79: Key challenges and opportunities in occupant-centric building design and operation. *Building and Environment*, 178, 106738.

Odling-Smee, F. J., Lala, K. N., & Feldman, M. (2003). *Niche Construction: The Neglected Process in Evolution (Monographs in Population Biology, 37)*. Princeton University Press.

OECD. (2018a). *Equity in Education: Breaking Down Barriers to Social Mobility*. OECD Publishing. https://doi.org/10.1787/9789264073234-en

OECD. (2018b). *Preparing our youth for an inclusive and sustainable world: The OECD PISA global competence framework* (ERIC No. ED581688). ERIC. https://eric.ed.gov/?id=ED581688

OECD. (2018c). *The future of education and skills: Education 2030* (OECD Education Policy Perspectives, No. 98). OECD Publishing. https://doi.org/10.1787/54ac7020-en

Okamura K., & Yamada S. (2020) Adaptive trust calibration for human-AI collaboration. *PLoS ONE* 15(2): e0229132. https://doi.org/10.1371/journal.pone.0229132

Olmsted, F. L. (1870). Public parks and the enlargement of towns. *Atlantic Monthly*, 25(150), 186–199.

Olson, G. M., & Olson, J. S. (2000). Distance matters. *Human–Computer Interaction*, 15(2), 139–178.

Oosterhuis, K., & Bier, H. H. (Eds.). (2013). *iA #5: Robotics in architecture* (124 pp.). Heijningen, The Netherlands: Jap Sam Books. ISBN 978-94-90322-31-1.

Oshinsky, D. M. (2005). *Polio: An American Story*. Oxford University Press.

Paglen, T. (2012). *The Last Pictures*. Creative Time Books.

Pallasmaa, J. (2012). *The Eyes of the Skin: Architecture and the Senses* (3rd ed.). John Wiley & Sons.

Pallasmaa, J. (2014). Space, place and atmosphere: Emotion and peripheral perception in architectural experience. *Lebenswelt*, 4(1), 230–245.

Parasuraman, R. (2011). Neuroergonomics: Brain, cognition, and performance at work. *Current Directions in Psychological Science*, 20(3), 181–186. https://doi.org/10.1177/0963721411409176

Parasuraman, R., & Riley, V. (1997). Humans and automation: Use, misuse, disuse, abuse. *Human Factors*, 39(2), 230–253.

Parasuraman, R., & Rizzo, M. (Eds.). (2007). *Neuroergonomics: The brain at work*. Oxford University Press.

Parasuraman, R., & Rizzo, M. (Eds.). (2009). *Neuroergonomics: The Brain at Work*, Human Technology Interaction Series (New York, 2006; online edn, Oxford Academic, 1 May 2009). Oxford University Press. https://doi.org/10.1093/acprof:oso/9780195177619.001.0001.

Parasuraman, R., & Wickens, C. D. (2008). Humans: Still vital after all these years of automation. *Human Factors*, 50(3), 511–520. https://doi.org/10.1518/001872008X312198

Parasuraman, R., Sheridan, T. B., & Wickens, C. D. (2000). A model for types and levels of human interaction with automation. *IEEE Transactions on Systems, Man, and Cybernetics - Part A: Systems and Humans*, 30(3), 286–297. https://doi.org/10.1109/3468.844354

Pariser, E. (2011). *The Filter Bubble: What the Internet Is Hiding from You*. Penguin Press.

Park, J. Y., Dougherty, T., Fritz, H., & Nagy, Z. (2019). LightLearn: An adaptive and occupant centered controller for lighting based on reinforcement learning. *Building and Environment*, 147, 397–414.

Pask, G. (1969). The architectural relevance of cybernetics. *Architectural Design*, 39(9), 494–496.

PassiveLogic. (2021). *Autonomous Buildings: Levels of Autonomy*. PassiveLogic White Paper.

Pedersen, K., Emblemsvag, J., Bailey, R., Allen, J., & Mistree, F. (2000). Validating design methods & research: The validation square. *Proceedings of DETC '00 2000 ASME Design Engineering Technical Conference*, Baltimore, MD.

Peffers, K., Tuunanen, T., Rothenberger, M. A., & Chatterjee, S. (2007). A design science research methodology for information systems research. *Journal of Management Information Systems*, 24(3), 45–77.

Pencavel, J. (2014). *The Productivity Puzzle: Restoring Economic Dynamism*. Princeton University Press.

Perdue, W. C., Stone, L. A., & Gostin, L. O. (2003). The built environment and its relationship to the public's health: The legal framework. *American Journal of Public Health*, 93(9), 1390–1394. https://doi.org/10.2105/AJPH.93.9.1390

Pérez-Gómez, A. (2016). *Attunement: Architectural Meaning after the Crisis of Modern Science*. MIT Press.

Perkins & Will. (2020). *Designing for Distance: Strategies for a Safe Workplace Re-entry*. Perkins & Will Research Journal.

Pflanzer, M., Traylor, Z., Lyons, J. B., Dubljević, V., & Nam, C. S. (2022). Ethics in human–AI teaming: Principles and perspectives. *AI and Ethics*, 3(3), 917–935.

Phuong, M., Zimmermann, R. S., Wang, Z., Lindner, D., Krakovna, V., Cogan, S., … & Shah, R. (2025). Evaluating Frontier Models for Stealth and Situational Awareness. arXiv preprint arXiv:2505.01420.

Picard, R. W. (1997). *Affective Computing*. MIT Press.

Polanyi, K. (1944). *The Great Transformation*. Rinehart.

Pollan, M. (2006). *The Omnivore's Dilemma: A Natural History of Four Meals*. Penguin Press.

Postman, N. (1985). *Amusing Ourselves to Death*. Viking.

Pouwels, A. (2020). *Open Plan Offices - The New Ways of Working*. European Parliament Briefing.

Quinones, S. (2015). *Dreamland: The True Tale of America's Opiate Epidemic*. Bloomsbury Press.

Radner, H. (2022). Anxiety in the concrete Jungle. *Psychology and Built Environments*, 6(1), 77–93.

Rajkumar, R., Lee, I., Sha, L., & Stankovic, J. (2010). Cyber-physical systems: The next computing revolution. *Proceedings of the 47th Design Automation Conference*, 731–736. https://doi.org/10.1145/1837274.1837461

Randall, T. (2015). *The Smartest Building in the World*. Bloomberg Businessweek.

Ranney, M. L., Griffeth, V., & Jha, A. K. (2020). Critical supply shortages—the need for ventilators and personal protective equipment during the COVID-19 pandemic. *New England Journal of Medicine*, 382(18), e41.

Rasmussen, J. (1983). Skills, rules, and knowledge; signals, signs, and symbols, and other distinctions in human performance models. *IEEE Transactions on Systems, Man, and Cybernetics*, 13(3), 257–266. https://doi.org/10.1109/TSMC.1983.6313160

Rasmussen, J., & Goodstein, L. P. (1985). Decision support in supervisory control. *IFAC Proceedings Volumes*, 18(10), 79–90. https://doi.org/10.1016/B978-0-08-032566-8.50015-1

Rasmussen, J., & Lind, M. (1981). *Coping with Complexity*. Risø National Laboratory. Risø-M No. 2293

Rasmussen, J., Pejtersen, A. M., & Schmidt, K. (1990). *Taxonomy for Cognitive Work Analysis*. Risø National Laboratory.

Ratti, C. (2016, June 6). *Carlo Ratti Outfits Open-Office with Individualized Climate and Lighting System*. designboom.

Ratti, C., & Claudel, M. (2015). *Open Source Architecture*. Thames & Hudson.

Ratti, C., & Claudel, M. (2016). *The City of Tomorrow: Sensors, Networks, Hackers, and the Future of Urban Life.* Yale University Press.

Ravani, S. (2019). *Oakland Bans Use of Facial Recognition Technology, Citing Bias Concerns.* San Francisco Chronicle.

Reed, B. (2007). Shifting from 'sustainability' to regeneration. *Building Research & Information*, 35(6), 674–680.

Reiche, D. (2010). Renewable energy policies in the gulf countries: A case study of Masdar City in Abu Dhabi. *Energy Policy*, 38(1), 378–382.

Reiche, D., & Mischke, P. (2019). Rethinking the future of low-carbon cities: Carbon neutrality, green design, and sustainability tensions in Masdar City. *Urban Planning*, 4(4), 146–157.

Robinson, W. (2020). Vaccine nationalism and the global health crisis. *Globalization and Health*, 16(1), 95–99.

Roosegaarde, D. (2011). *Dune 4.1–4.2. Studio Roosegaarde [Interactive artwork].*

Rosen, L. D. (2016). *The Distracted Mind: Ancient Brains in a High-Tech World.* MIT Press.

Rosen, L. D. (2017). *The Distracted Student Mind: Enhancing Its Focus and Attention.* Phi Delta Kappan.

Ross, J. W., Weill, P., & Robertson, D. C. (2006). *Enterprise Architecture as Strategy: Creating a Foundation for Business Execution.* Harvard Business School Press.

Ruiz, J., Li, Y., & Lank, E. (2011). User-defined motion gestures for mobile interaction. *Proceedings of the SIGCHI Conference on Human Factors in Computing Systems* (pp. 197–206). ACM.

Sadowski, J. (2020). *Too Smart: How Digital Capitalism is Extracting Data, Controlling Our Lives, and Taking Over the World.* MIT Press.

Sadowski, J., & Bendor, R. (2019). Selling smartness: Corporate narratives and the smart city as a sociotechnical imaginary. *Science, Technology, & Human Values*, 44(3), 540–563. https://doi.org/10.1177/0162243918806061

Said, E. W. (1978). *Orientalism.* Pantheon Books.

Salama, A. M. (2015). *Spatial Design Education: New Directions for Pedagogy in Architecture and Beyond.* Routledge.

Sallis, J. F., Floyd, M. F., Rodríguez, D. A., & Saelens, B. E. (2012). Role of built environments in physical activity, obesity, and cardiovascular disease. *Circulation*, 125(5), 729–737. https://doi.org/10.1161/CIRCULATIONAHA.110.969022

Salonen, H., Lahtinen, M., & Lappalainen, S. (2020). Post-occupancy evaluation of workplace environments: An argument for evidence-based workplace management. *Intelligent Buildings International*, 12(2), 101–114.

Salvendy, G. (2012). *Handbook of Human Factors and Ergonomics* (4th ed.). Wiley.

SAMHSA. (2014). *SAMHSA's Concept of Trauma and Guidance for a Trauma-Informed Approach.* Substance Abuse and Mental Health Services Administration.

Sandel, M. J. (2007). *The Case Against Perfection: Ethics in the Age of Genetic Engineering.* Harvard University Press.

Sanders, E. B.-N., & Stappers, P. J. (2008). Co-creation and the new landscapes of design. *CoDesign*, 4(1), 5–18.

Sanoff, H. (2000). *Community Participation Methods in Design and Planning.* Wiley.

Santamouris, M., Cartalis, C., Synnefa, A., & Kolokotsa, D. (2015). On the impact of urban heat island and global warming on the power demand and electricity consumption of buildings—A review. *Energy and Buildings*, 98, 119–124, https://doi.org/10.1016/j.enbuild.2014.09.052.

Sapolsky, R. M. (2004). *Why Zebras Don't Get Ulcers.* Holt Paperbacks.

Sassen, S. (2017). Predatory formations dressed in wall street suits and algorithmic math. *Science, Technology & Society*, 22(1), 6–20. https://doi.org/10.1177/0971721816682783

Sauda, E., Karduni, A., & Lanclos, D. (2024). *Architecture in the Age of Human–Computer Interaction.* Routledge.

Scharff, R. C., & Dusek, V. (Eds.). (2003). *Philosophy of technology: The technological condition—An anthology* (1st ed.). John Wiley & Sons.

Schneider, T., & Till, J. (2007). *Flexible Housing.* Architectural Press.

Scott, C. F., Marcu, G., Anderson, R. E., Newman, M. W., & Schoenebeck, S. (2023). Trauma-informed social media: Towards solutions for reducing and healing online harm. In *Proceedings of the 2023 CHI Conference on Human Factors in Computing Systems* (pp. 1–20).

Sennett, R. (2007). *The Culture of the New Capitalism.* Yale University Press.

Sennett, R. (2013). *Together: The Rituals, Pleasures and Politics of Cooperation.* Yale University Press.

Sennett, R. (2018). *Building and Dwelling: Ethics for the City.* Farrar, Straus and Giroux.

Shahriari, K. & Shahriari, M. (2017). IEEE standard review — Ethically aligned design: A vision for prioritizing human wellbeing with artificial intelligence and autonomous systems. *2017 IEEE Canada International Humanitarian Technology Conference (IHTC),* Toronto, ON, Canada, pp. 197–201. https://doi.org/10.1109/IHTC.2017.8058187

Shiva, V. (2015). *Earth Democracy: Justice, Sustainability, and Peace.* South End Press.

Shneiderman, B. (2020). Human-centered artificial intelligence: Three fresh ideas. *AIS Transactions on Human-Computer Interaction,* 12(3), 109–124. https://doi.org/10.17705/1thci.00131

Shuchman, M. (2014). Ebola outbreak exposes the potential—and limitations—of mobile health technology. *Canadian Medical Association Journal,* 186(15), 1143.

Sicari, S., Rizzardi, A., Grieco, L. A., & Coen-Porisini, A. (2015). Security, privacy and trust in Internet of Things: The road ahead. *Computer Networks,* 76, 146–164.

Simon, H. A. (1971). Designing organizations for an information-rich world. In: M. Greenberger (ed.), *Computers, Communications, and the Public Interest* (pp. 37–72). Johns Hopkins Press.

Simon, M. (2024). City trees save lives. *Wired Magazine Article.*

Sirkin, D., Mok, B., Phan, L., & Ju, W. (2015). Mechanical ottoman: How robotic furniture offers and withdraws support. *Proceedings of the 10th ACM/IEEE International Conference on Human-Robot Interaction* (pp. 11–18). ACM.

Smart, P. R. (2017). Extended cognition and the Internet: A review of current issues and controversies. *Philosophy & Technology,* 30(3), 357–390. https://doi.org/10.1007/s13347-016-0250-2

Smith, A., & Jones, B. (2019). Digital marketing strategies and consumer behavior: Insights from the retail sector. *Journal of Emerging Trends in Marketing,* 7(1), 45–56.

Smith, C., & Rowe, G. (2016). Deliberative Processes in Practice. In: Dodds, S., Ankeny, R. (eds) *Big Picture Bioethics: Developing Democratic Policy in Contested Domains. The International Library of Ethics, Law and Technology,* vol. 16. Springer, Cham. https://doi.org/10.1007/978-3-319-32240-7_5

Snow, J. (1855). *On the Mode of Communication of Cholera.* John Churchill.

Snowden, F. M. (2019). *Epidemics and Society: From the Black Death to the Present.* Yale University Press.

Society of Automotive Engineers (SAE). (2021). International recommended practice, taxonomy and definitions for terms related to driving automation systems for on-road motor vehicles, SAE Standard J3016_202104, Revised April 2021, Issued January 2014. https://doi.org/10.4271/J3016_202104.

Sommerville, I., & Sawyer, P. (1997). *Requirements Engineering: A Good Practice Guide.* John Wiley & Sons, Inc.

Spinuzzi, C. (2012). Working alone together: Coworking as emergent collaborative activity. *Journal of Business and Technical Communication,* 26(4), 399–441.

Spivak, G. C. (1988). Can the subaltern speak? In C. Nelson & L. Grossberg (Eds.), *Marxism and the Interpretation of Culture* (pp. 271–313). University of Illinois Press.

Standing, G. (2011). *The Precariat: The New Dangerous Class.* Bloomsbury Academic.

Stanford University. (2019). Sustainability at Stanford: 2018–2019 year in review. Retrieved from https://sustainability-year-in-review.stanford.edu/2019/

Stanford University News. (2019). Stanford opens 'team science' complex for brain research and molecular discovery. *Stanford News*, November 19, 2019.

Stanton, N. A. (2006). Hierarchical task analysis: Developments, applications, and extensions. *Applied ergonomics*, 37(1), 55–79.

Stephens, A., & Patel, S. (2021). Rethinking insulative materials in the era of climate extremes. *Journal of Urban Infrastructure*, 14(2), 213–228.

Sternberg, Esther M. (2010). *Healing Spaces: The Science of Place and Well-Being* (2nd ed.). Belknap Press: An Imprint of Harvard University Press.

Steyerl, H. (2013). *How Not to Be Seen: A Fucking Didactic Educational. MOV File [Video Installation]*. Venice Biennale, Italy.

Stokols, D., Lejano, R. P., & Hipp, J. (2013). Enhancing the resilience of human–environment systems: A social–ecological perspective. *Ecology and Society*, 18(1), 7.

Sullivan, C., & Petrovich, L. (2021). Urban forests: A missing link in climate resilience. *Environmental Psychology Quarterly*, 9(3), 212–228.

Sundstrom, E., Town, J. P., Rice, R. W., Osborn, D. P., & Brill, M. (1994). Office noise, satisfaction, and performance. *Environment and Behavior*, 26(2), 195–222. https://doi.org/10.1177/001391659402600204

Sunstein, C. (2009). *Republic.com 2.0*. Princeton University Press.

Susskind, R., & Susskind, D. (2022). *The Future of the Professions: How Technology Will Transform the Work of Human Experts*. Oxford University Press.

Sussman, J. M. (2007). *The CLIOS process: A user's guide (Course materials for ESD.04J/1.041J Frameworks & Models in Engineering Systems, Spring 2007)*. Massachusetts Institute of Technology. Available from MIT OpenCourseWare.

Taleb, N. N. (2007). *The Black Swan: The Impact of the Highly Improbable*. Random House.

Tan, P. Y., Wang, J., & Sia, A. (2013). Perspectives on five decades of the urban greening of Singapore. *Cities*, 32, 24–32. https://doi.org/10.1016/j.cities.2013.02.001

Tapus, A., Mataric, M. J., & Scassellati, B. (2007). Socially assistive robotics [Grand challenges in robotics]. *IEEE Robotics & Automation Magazine*, 14(1), 35–42.

The Guardian. (2024). Prof Andrea Mechelli: 'People who live near green space are less likely to struggle with mental health issues'. *The Guardian Article*.

The Lancet. (2021). Global COVID-19 vaccine inequity. *The Lancet*, 398(10296), 1533.

The Open Group. (2018). *TOGAF® Version 9.2*. The Open Group.

Thunberg, G. (2019). *No One Is Too Small to Make a Difference*. Penguin Books.

Till, J. (2009). *Architecture depends*. MIT Press.

Time (2024). How cities are using nature-based solutions to tackle floods. *Time Magazine Article*.

Topol, E. (2019). The Topol Review: preparing the healthcare workforce to deliver the digital future: an independent report on behalf of the Secretary of State for Health and Social Care, February 2019, Health Education England, https://hdl.voced.edu.au/10707/687370.

Trojanová, M., Hošovský, A., & Čakurda, T. (2023). Evaluation of machine learning-based parsimonious models for static modeling of fluidic muscles in compliant mechanisms. *Mathematics*, 11(1), 149. https://doi.org/10.3390/math11010149

Troncoso, C., Payer, M., Hubaux, J.-P., Salathé, M., Larus, J., & Bugnion, E. (2020). *Decentralized Privacy-Preserving Proximity Tracing*. White Paper.

Tschumi, B. (1996). *Architecture and Disjunction*. MIT Press.

Tschumi, B. (2001). *Event-Cities 2*. MIT Press.

Tschumi, B. (2005). *Event-Cities 3: Concept vs. Context vs. Content*. MIT Press.

Tschumi, B. (2012). *Architecture Concepts: Red is Not a Color*. Rizzoli Press.

Turkle, S. (2011). *Alone Together: Why We Expect More from Technology and Less from Each Other*. Basic Books.

Turkle, S. (2015). *Reclaiming Conversation: The Power of Talk in a Digital Age.* Penguin Press.

Ulrich, R. S. (1984). View through a window may influence recovery from surgery. *Science*, 224(4647), 420–421.

Ulrich, R. S. (2001). Effects of healthcare environmental design on medical outcomes. In A. Dilani (Ed.), *Design and Health: Proceedings of the Second International Conference on Health and Design.* International Academy for Design and Health.

Ulrich, R. S., Simons, R. F., Losito, B. D., Fiorito, E., Miles, M. A., & Zelson, M. (1991). Stress recovery during exposure to natural and urban environments. *Journal of Environmental Psychology*, 11(3), 201–230.

Ulrich, R. S., Zimring, C., Joseph, A., Quan, X., & Choudhary, R. (2004). *The Role of the Physical Environment in the Hospital of the 21st Century: A Once-in-a-Lifetime Opportunity.* The Center for Health Design.

Ulrich, Roger, Zimring, Craig, Zhu, Xuemei, DuBose, Jennifer, Seo, Hyun, Choi, Young-Seon, Quan, Xiaobo, & Joseph, Anjali. (2008). A review of the research literature on evidence-based healthcare design. *HERD*, 1, 61–125. https://doi.org/10.1177/193758670800100306.

Ursprung, P. (Ed.). (2005). *Herzog & de Meuron: Natural History.* Lars Müller Publishers.

Vale, V. (2008). *The Reconfigurable House, How to Remain Smarter Than Your House.* Neural.it. https://neural.it/2008/06/the-reconfigurable-house-how-to-remain-smarter-than-your-house/

Van Dijck, J. (2014). Datafication, dataism and dataveillance: Big Data between scientific paradigm and ideology. *Surveillance & Society*, 12(2), 197–208. https://doi.org/10.24908/ss.v12i2.4776

Varela, F. J., Thompson, E., & Rosch, E. (1991). *The Embodied Mind: Cognitive Science and Human Experience.* MIT Press.

Vartanian, O., Navarrete, G., Chatterjee, A., Fich, L. B., Leder, H., Modroño, C., … Skov, M. (2013). Impact of contour on aesthetic judgments and approach-avoidance decisions in architecture. *Proceedings of the National Academy of Sciences*, 110, 10446–10453. https://doi.org/10.1073/pnas.1301227110

Verma, S., Gonthina, P., Hawks, Z., Nahar, D., Brooks, J. O., Walker, I. D., Wang, Y., de Aguiar, C., & Green, K. E. (2018). Design and evaluation of two robotic furnishings partnering with each other and their users to enable independent living. *Proceedings of the 12th EAI International Conference on Pervasive Computing Technologies for Healthcare (PervasiveHealth '18).* Association for Computing Machinery, New York, 35–44. https://doi.org/10.1145/3240925.3240978

Vicente, K. J. (1999). *Cognitive Work Analysis: Toward Safe, Productive, and Healthy Computer-Based Work.* Lawrence Erlbaum Associates.

Vischer, J. C. (2007). The effects of the physical environment on job performance: Towards a theoretical model of workspace stress. *Stress and Health*, 23(3), 175–184. https://doi.org/10.1002/smi.1134

Wallach, W., & Allen, C. (2009). *Moral Machines: Teaching Robots Right from Wrong.* Oxford University Press.

Wang, Y., Green, K. E. (2023). How do we want to interact with robotic environments? User preferences for embodied interactions from pushbuttons to AI. In P. Morel, & H. Bier (Eds.), *Disruptive Technologies: The Convergence of New Paradigms in Architecture. Springer Series in Adaptive Environments.* Springer. https://doi.org/10.1007/978-3-031-14160-7_3

Webster, H. (2008). Architectural education after Schön: Cracks, blurs, boundaries and beyond. *Journal for Education in the Built Environment*, 3(2), 63–74.

Weiser, M., & Brown, J. S. (1997). The coming age of calm technology. In P. Denning & R. Metcalfe (Eds.), *Beyond Calculation: The Next Fifty Years of Computing.* Springer.

WHO. (2018). *Noncommunicable Diseases Country Profiles.* World Health Organization.

WHO. (2019). *Burn-Out an "Occupational Phenomenon": International Classification of Diseases*. World Health Organization.

Wickens, C. D. (1981). The structure of attentional resources. In R. Nickerson (Ed.), *Attention and Performance VIII* (pp. 239–257). Psychology Press.

Wickens, C. D. (1984). Processing resources in attention. In R. Parasuraman & D. R. Davies (Eds.), *Varieties of attention* (pp. 63–102). Academic Press.

Wickens, C. D. (2002). Multiple resources and performance prediction. *Theoretical Issues in Ergonomics Science*, 3(2), 159–177. https://doi.org/10.1080/14639220210123806

Wickens, C. D., & Hollands, J. G. (1999). *Engineering Psychology and Human Performance* (3rd ed.). Pearson.

Wiegers, K., & Beatty, J. (2013). *Software Requirements* (3rd ed.). Microsoft Press.

Wiener, N. (1954). *The human use of human beings: Cybernetics and society* (Revised ed.). Houghton Mifflin.

Wiener, N. (1961). *Cybernetics: Or Control and Communication in the Animal and the Machine*. MIT Press.

Willis, K. S., & Aurigi, A. (2017). *Digital and Smart Cities*. Routledge.

Wilson, M. (2002). Six views of embodied cognition. *Psychonomic Bulletin & Review*, 9(4), 625–636.

Wood, A. J., Graham, M., Lehdonvirta, V., & Hjorth, I. (2019). Networked but Commodified: The (Dis)Embeddedness of Digital Labour in the Gig Economy. *Sociology*, 53(5), 931–950. https://doi.org/10.1177/0038038519828906

World Economic Forum. (2020). *The Future of Jobs Report 2020*. World Economic Forum.

World Health Organization. (2018). *Noncommunicable Diseases Country Profiles*. WHO Press.

World Health Organization. (2019). Burn-out an "occupational phenomenon": International Classification of Diseases. https://www.who.int/mental_health/evidence/burn-out/en/

World Health Organization. (2020). *Coronavirus Disease (COVID-19) Pandemic*. WHO Official Website.

World Health Organization. (2021). *Mental Disorders*. WHO Official Website.

World Health Organization (WHO). (2021). An unwavering voice for vaccine equity. from the WHO results report, 2020–2021.

Wu, H.-K., Lee, S. W.-Y., Chang, H.-Y., & Liang, J.-C. (2013). Current status, opportunities and challenges of augmented reality in education. *Computers & Education*, 62, 41–49.

WWF. (2020). *Living Planet Report 2020*. World Wildlife Fund.

Xie, F. (2006). "Component-Based Cyber-Physical Systems. in *NSF Workshop on CyberPhysical Systems*, Austin, TX.

Yao, L., Ou, J., Cheng, C. Y., et al. (2013). PneUI: Pneumatically actuated soft composite materials for shape changing interfaces. *Proceedings of the 26th Annual ACM Symposium on User Interface Software and Technology* (pp. 13–22). ACM.

Young, R. (2004). *The Requirements Engineering Handbook*. Artech House.

Young, M. S., Brookhuis, K. A., Wickens, C. D., & Hancock, P. A. (2014). State of science: Mental workload in ergonomics. *Ergonomics*, 58(1), 1–17.

Zachman, J. A. (1987). A Framework for Information Systems Architecture. *IBM Systems Journal*, 26(3), 276–292.

Zeisel, J. (2006). *Inquiry by Design: Environment/Behavior/Neuroscience in Architecture, Interiors, Landscape, and Planning* (Rev. ed.). W. W. Norton & Company.

Zuboff, S. (2019). *The Age of Surveillance Capitalism: The Fight for a Human Future at the New Frontier of Power*. PublicAffairs.

Index

Pages in *italics* refer to figures and pages in **bold** refer to tables and pages in ***bold italics*** refers to box.

A

adaptive environments, 1, 5–7, 9–11, 96–97, 102, 122–123, 130, 150, 153
agent, 4–9, 11, 15, 22, 52, 93, 95–96, 109, 111, 122–125, 129, 132–134, 149–151, 154–156, 159, 161, 164, *166*, 171–172
All-in-One (AIO) Environments, 10–11, 151–174
allocentric, mutualistic symbiosis, 6–7, 124–125, 132–133, 150, 156
Andy Clark, 3, 107, 113, 130, 144, 149, 153–154, 159
Arendt, 17
artifact ecologies, 11, 149
Artificial Intelligence (AI), 1, 4, 6–10, 16, 19, 21, 47, 49, 73–74, ***75–77, 79***, 93, 95–96, 103–104, 112–113, 117, 123–126, 128–132, 136, 139, 153–157, 159–160, 162–164, ***166, 169***, 174
automation, 1–2, 7–8, 15–17, 25, 73–74, 91, 95–96, 103, 122, 131–132, 155, 158, 162
Autonomous Driving System (ADS), 7–8, 10, 16, *57*, 86, 96
autonomy, xiii, 16, 26–27, **31, 34**, 40, 43–45, 47, 60, 95–96, 99, 113, 116, 126, 131, 136, 155, 158–159, 162, 164, 174

B

biophilic, 46–47, 55, 60, 88, 129, *140*, 157, 171
Building Information Model (BIM), 49, 96

C

calibrated trust, 6, 103–104, 130–132, 155, 159, 162
calm technology, 117, 154
Carr, 24, 26, 43–44, 54, 69
Churchill, 5
coevolution, 4–5, 7, 10–11, 122–123, 152, 171
coexistence, 4–5, 7, 10–11, 29, 122–123, 125, 152, 160, 171
cognition, xii–xiii, 2–4, 6, 10, 17, 95–96, 98, 106–113, 122–124, 130, 144, 149–150, 154–155, 159, 164, *168*
Cognitive Work Analysis (CWA), 101, *102, 110*

colonization, xii–xiii, 5, 10, 12–14, 17, 19, 22–28, 30, 37, 39
  digital and cyber colonization, 24–25
  environmental and ecological, 27–29
  genetic developments, 23, 42
  health colonization, 25–27
  pandemics and viruses, 22
Complex, Large-Scale, Integrated, Open, Sociotechnical Systems (CLIOS), 94–95
Control Task Analysis (CTA), 101
cortisol, 8, 60, 69, 88
COVID-19, 22, 39, 41, 43–44, 50, 65, 88
Cross-Programming, 6, 88, 115–117, 122
Cyber-Physical Systems (CPS), 9, 50, 90–96, 113, 130, 139, 145, 155

D

Deleuze, 14, 17, 53, 88, 130, 148
Derrida, 148

E

Ecological-Niche Construction (ENC), 4, 6, 51, 111, 123–124, 129, 136, 148–150, 153
egocentric, 124
Eliot, 14
Embodied Cognition, xiii, 3, 6, 10, 106–107, 122–124, 130, 144, 149–150, 154, 159, *168*
ethical, 1, 4, 19, 23, 40, 44, 89–90, 104, 112–113, 122, 127–133, 136, 139, 148–149, 153, 162–164, 174
Evidence-Based Design (EBD), 4, 6, 39, 48, 51, 65, 88, 100, 102, 105, 111, 123, 129, 154
Extended Reality (XR), 10

F

Foucault, 15–17, 21
Fundamental Question, 1–6, 153

G

Guattari, 14, 53, 130

H

health, 7–10, 15–16, 21–22, 25–29, **31**, **34–35**,
    39–47, 54–55, *57*, 59–60, 64–65, *66–68*,
    70, 81, 88–90, 99, 120, 129, 141–142, 145,
    148, 155, 158, *168*, 170
healthcare, xii, 18, 26, 30, **34**, 39, 41, 44–45,
    48, 52, 54–55, 64–65, *66*, *68*, 103, 112,
    139–142
Hedonic Psychology, 98–99, 122, 139
Heidegger, 14, 30, 147–148
heroes, 10–11, 37–41, 43–47, 49, 52, 86–87, 122
Human-AI-Robot Teaming (HART), 6–7, 9, 95,
    109, 122, 124, 130, 133, 150, 153–154,
    156–157, 159
Human-Computer Interaction (HCI), 104, 110,
    112, 131
Human Readiness Levels (HRL), 96

I

Internet-of-Things, 1, **34**, 90
Internet-of-Things-Enabled Smart Cities (IES), 92
Industry 3.0, 1–2
Industry 4.0, 1–4, 11, 15, 96–97
Industry 5.0, 2–3, 96–97
Industry 6.0, 1–6, 10–11, 87, 89, 94, 96–97, 115,
    122, 149, 153
integrated, xiii, 4–5, 9, 11, 14, 21, **35**, 39, 42, 48,
    56, *61*, 65, 85, 90, *91*, 92, 94, 99, 105,
    110–113, **116**, 121, 129, 131, 145, 156,
    159
integration, 2, **34–36**, 43, 49, 54–55, 60, 65,
    67–68, 90–93, 95–97, 100, 111–113, 117,
    121–122, 126, 129–131, 135–136, 139,
    144–146, 150, 154, 156–157, 159, 161
integrity, 6–7, 23, *57*, *62*, *77*, 95, 125, 132–133,
    155–156
intelligent, xii–xiii, 1, 3, 5–7, 9–11, 88–90,
    96–97, 100, 102–103, *106*, 119, 121–124,
    126, 128, 130–134, 136, 147, 149–154,
    159–164
interaction, 1, 11–12, 14, 19, 21, 24–25, **32–33**,
    35, 43, 48, 54, *58–59*, 78, 85, 90, 95–96,
    99, 101, 104, 107–108, 111–115, **116–117**,
    117–121, 125, 127, 130–131, *135*, 141,
    143–146, 155–157, 159–160, 162, *168*,
    170–171
interactive, xiii, 4, 20–21, 39–40, 48, 86, 88, 90,
    102–104, 112, 115, 117, 119–122, 131,
    133, 145, *146*, 150, 152, 154, 159, *167*,
    169–170
International Space Station, 3, 88

J

Jacobs, 14, 39, 43, 51, 125, 127

K

Kirsh, 9, 111–112, 149, 153–156

L

learning, 19, **33–34**, 73–74, *77*, 86, 88, 104,
    109–110, 119, 134, 136, 142, 153–155,
    158, 161, 163, *167–168*, 170, 173–174
levels of automation, 95–96, 131
lighting, 26, 43, 64–65, *66*, *70–71*, *77*, 81, *83*, 86,
    88, 90, 93, 95, 98–99, 107, **116**, 119–120,
    124, 130, 135, 142, 148, 150, 154, 158–161,
    165–166, *169–170*, 172–173
living, 2, 7, 24, 27–29, **32**, 39–40, 42, 45–46,
    49, *62*, *68*, *79*, 82, *84*, 85, 88, 117,
    119–121, 123, 125–127, 129, 134, 137,
    141, 143–145, 148–149, 153, 155–157,
    162–163, *166*, *168–170*, 171, 174

M

Machine learning (ML), 104, 136, 154–155, 161
Manganelli, 1, 90–91, *92*, 102–103, *106*,
    109–110, 111, 117, 136, 145, 149, 154,
    156, 159
Masdar City, 126, 136–139
Mental model, 2, 11, 105, 160, 162
Mokhtar, 117–121, 129, 131, 136, 145, 148–149,
    154–155, 159, 161
Multiple Resource Theory (MRT), 98–99, 157
mutualistic symbiosis, 6–7, 124–125

N

neuroadaptive, 106–107, 131, 150, 157
Neuroergonomics, 17, 106–107, 150, 153, 155
Non-Humanoid Social Robotics (NH-SR), 6,
    117–122, 126, 129, 131, 146, 148–150,
    153–155, 161–162

O

Odling-Smee, 4, 51, 55, 149
Operating Envelope Specification (OES), 6, 96
Operational Design Domain (ODD), 6, 96

P

Parasuraman, 17, 104, 107, 112, 131, 146, 155–156
Poly-Colonization, 13, 15–21, 23–25, 27, 29–30, 46
Psychofortology, 97, 99, 139, 157

R

Requirements Engineering (RE), 6, 99–101,
    122, 129

robotics, 1, 4, 6, 9–10, 73, *76*, 117, 119–121,
    131–132, 145–147, 155–156, 158–159,
    161–162

S

sensing, xii, 2, 6, 88, 93, 95, 110, 119, 130,
    155–156, 160, 172
sensor, 2–3, 13–14, 21, 25, **33**, **35**, 46, 48–49,
    54–56, *57*, 60, *66*, 74, 89–91, 96, 117, 119,
    121, 130, 132, 135–136, 145–147, 154,
    156–161, 163, *169*
smart, 4, 9–10, 14, 16, 23, 25–26, **31**, 40, 43, 53,
    55–56, *58*, 90, 92–93, 102, 113, 119–120,
    128, 130–132, 135–136, 152, 160–161, 164
Society of Automotive Engineers (SAE), 95–96,
    155, 158
Sociotechnical Systems (STS), 6, 10, *92*, 93–95,
    101, 105, 110, 115, 127, 130, 155
Stanford Neurosciences Institute (SNI), 142–144
stress, xii–xiii, 8, 11, 17, 26–27, **32**, **34–36**, 38,
    46–47, 54, 56, 59–60, *62–64*, 65–66, *67*,
    69–70, 81, 87–90, 99, 102, 104, 107, 120,
    136, 139, 148, 150, 152, 156, 158, 161,
    171–172
stressor, xii–xiii, 7, 9–10, 29–30, **31–32**, **34**, **36**,
    37, 40, 45, 53, 60, *66*, 87, 145
Super-Imposition, 115–116, **116–117**, 122
surveillance, 9, 13, 16, 20, 22, 26, 30, **31–35**, 42,
    47, 52–57, *58*, *80*, 87, 104, 125, 152
surveillance capitalism, 15, 17, 47
symbiosis, 6–7, 11, 93, 123–129, 134, 136–138,
    141, 144, 148, 154, 164, 170, 173–174
symbiotic, xiii, 6–7, 11, 124–129, 132–134,
    136–139, 141–142, 144–146, 149–150,
    154, 156, 164, *168*, 171, 173
Systems Modeling Language (SysML), 102–103

T

The Edge, 90, 130, 135–136, *135*
Trans-Programming, 6, 39, 113–115, **116–117**,
    122, 149

trauma, xii–xiii, 4–11, 20–21, 29–30, **31–36**,
    37, 52, 54–55, 60, 88, 97–99, 122, 139,
    141–142, 145, 148–149, 151–152, 171
Trauma-Informed Design (TID), 10, 88, 90,
    98–99, 122, 139, 148, 153, 156, 158
trust, 6–9, 15, 18, 21, 27, **33**, 43, 93, 99, 103–104,
    113, 124–125, 128, 130–133, 155–160,
    162–164, 173–174
trustworthiness, 6–7, 14, 101, 103, 125, 132–133,
    155–156
Tschumi, 14, 39, 88, 113–117, **116–117**, 130,
    145, 164
Turkle, 24, 26, 42, 54, 113

U

Ubiquitous Computing, 117, 130, 154
Ulrich, 39, 41–42, 46, 48, 64–65, 88, 112
Ultra-Large-Scale Systems (ULSS), 89–90, *92*,
    94–95, 97, 109–110, 112–113, 115, 122,
    126, 128, 130, 133, 149–150, 153–155,
    159

V

Virtual Reality (VR), 73, *76*, 82
virtue, 6–7, 51, 64, 125, 132–133, 155–156

W

well-being, xii, 1, 9, 26–28, 37, 40, 47, 50, 54–55,
    68, 86, 88, 90, 93, 97–99, 103, 107, 110,
    112–113, 117, 128–129, 131, 141–142,
    144–146, 148–149, 154–157, *168*, 173
Work Domain Analysis (WDA), 101
World Health Organization (WHO), 22, 26, 42,
    44–45, 70

Z

Zuboff, 15, 17, 24, 43, 47, 56

For Product Safety Concerns and Information please contact our EU
representative GPSR@taylorandfrancis.com
Taylor & Francis Verlag GmbH, Kaufingerstraße 24, 80331 München, Germany